BURLEIGH DODDS SCIENCE: INSTANT INSIGHTS

NUMBER 04

Nutraceuticals in fruit and vegetables

T0176452

burleigh dodds
SCIENCE PUBLISHING

Published by Burleigh Dodds Science Publishing Limited
82 High Street, Sawston, Cambridge CB22 3HJ, UK
www.bdspublishing.com

Burleigh Dodds Science Publishing, 1518 Walnut Street, Suite 900, Philadelphia, PA 19102-3406, USA

First published 2021 by Burleigh Dodds Science Publishing Limited

British Library Cataloguing in Publication Data
A catalogue record for this book is available from the British Library

ISBN 978-1-78676-924-4 (Print)
ISBN 978-1-78676-925-1 (ePub)

DOI 10.19103.9781786769251

Typeset by Deanta Global Publishing Services, Dublin, Ireland

Contents

Chapter 1

Bioactive/nutraceutical compounds in fruit that optimize human health benefits

Federica Blando and Miriana Durante, Institute of Sciences of Food Production (ISPA), Italy; and B. Dave Oomah, formerly Pacific Agri-Food Research Centre, Canada

1   Introduction

2   Polyphenols

3   Carotenoids

4   Vitamin C

5   Production practices that influence bioactive compound synthesis

6   Future trends and conclusion

7   Where to look for further information

8   References

1 Introduction

In the past decade, numerous reports have demonstrated that high consumption of fruits and vegetables is beneficial for health, being associated with a reduced risk of degenerative diseases (Miller et al., 2017). Fruits are an important part of the human diet and a major source of bioactive compounds (BCs) with human health benefits. Unfortunately, this 'important part' of the human diet has been a decreasing proportion of the Western diet for the last 50 years. The 'Five a Day' program was introduced by national campaigns in several Western countries according to World Health Organization (WHO) and United Nations Food and Agriculture Organization (FAO) recommendations in their joint initiative (launched in 2003) to promote fruits and vegetables for health worldwide. This program has been implemented to better communicate updated dietary guidelines, which recommend more than five servings of fruits and vegetables. A recent English study found that eating seven or more portions of fruits and vegetables reduces the specific risks of death by cancer and heart disease (25% and 31%, respectively) (Oyebode et al., 2013).

This chapter provides a brief description of the chemistry of BCs and their presence in (temperate) fruits, and discusses recent advances in strategies towards improving sustainable crop production for nutraceuticals.

Among different BCs present in fruits are polyphenols, carotenoids and some vitamins (C), which are all correlated with antioxidant potential of fruits. The content of BCs in

http://dx.doi.org/10.19103/AS.2018.0040.14

fruits can vary depending on different factors, primarily genetics and the environmental conditions where fruit trees are cultivated. In the last 30 years, the genetic improvement of fruit trees has mainly been driven by productivity issues and consumer preferences; fruit quality traits have only recently been recognized as important additions to production traits (size, high productivity etc.). The nutraceutical characterization of fruits has become an additional factor to consider for genetic improvement programmes. Unfortunately, the selective pressure for commercial traits has resulted in a reduction of BC content in most modern cultivars, relative to older varieties. For this reason, the preservation and valorization of fruit germplasm recently became a key issue, with the re-discovery of ancient varieties strongly linked with their territory of origin. This has become a leading trend in modern fruit breeding and cultivation, at least in Europe.

2 Polyphenols

2.1 Polyphenols in general

Phenolics, one of the most diverse plant food antioxidants, are characteristic features of all plant tissues, forming an extensive group of compounds with a common hydroxy-substituted benzene ring structure. Chemically, phenols are reactive substances, very susceptible to oxidation (leading to quinones during storage or processing of plant foods), but they are also very good antioxidants, thanks to hydroxyl groups, especially at the *ortho*-position (Harborne, 1980; Parr and Bolwell, 2000).

Phenolics are mostly synthesized from phenylalanine through phenylalanine ammonia-lyase (PAL) and further metabolism of the simple C6-C3 phenylpropanoids.

Most often, phenolics are grouped into non-flavonoid (hydroxybenzoic acid and hydroxycinnamic acid derivatives) and flavonoid classes (Fig. 1). The phenolics hydroxycinnamic and chlorogenic acids predominate in the non-flavonoid phenolic class, whereas anthocyanins represent the most common sub-class of the flavonoid group. Other investigators divide phenolic compounds into sixteen classes based on their molecular structure, with flavonoids and phenolic acids most prevalent in fruits (Manganaris et al., 2014).

Phenolics have been associated with health benefits due to their antioxidant capacity, which varies with the chemical structure. A broad definition of antioxidant is 'any substance that, when present at low concentrations compared to those of an oxidizable substrate, significantly delays or prevents oxidation of that substrate' (Halliwell, 1994). Dealing with antioxidant properties and their relation with biological systems, it is important to underline that the antioxidant compound needs to remain stable after its reducing activities have been exerted.

Phenolics can act as antioxidants in a number of ways: chelating metals responsible for non-enzymatic free radical generation, breaking the cycle of generating new radicals or inhibiting radical generation from pro-oxidant enzymes. Phenolics are able to act not only as antioxidants, but they have shown anticarcinogenic action, in many studies, both blocking/suppressing carcinogenesis, inhibiting enzymes, and showing specific receptor interactions (Parr and Bolwell, 2000).

The antioxidant capacity of fruit extracts can be assayed *in vitro* by different tests, based on the scavenging activity of the BC against different radicals, such as the hydroxyl radical in the Trolox Equivalent Antioxidant Capacity assay (TEAC), and the peroxyl radical in the Oxygen Radical Absorbance Capacity assay (ORAC), and others. Total antioxidant

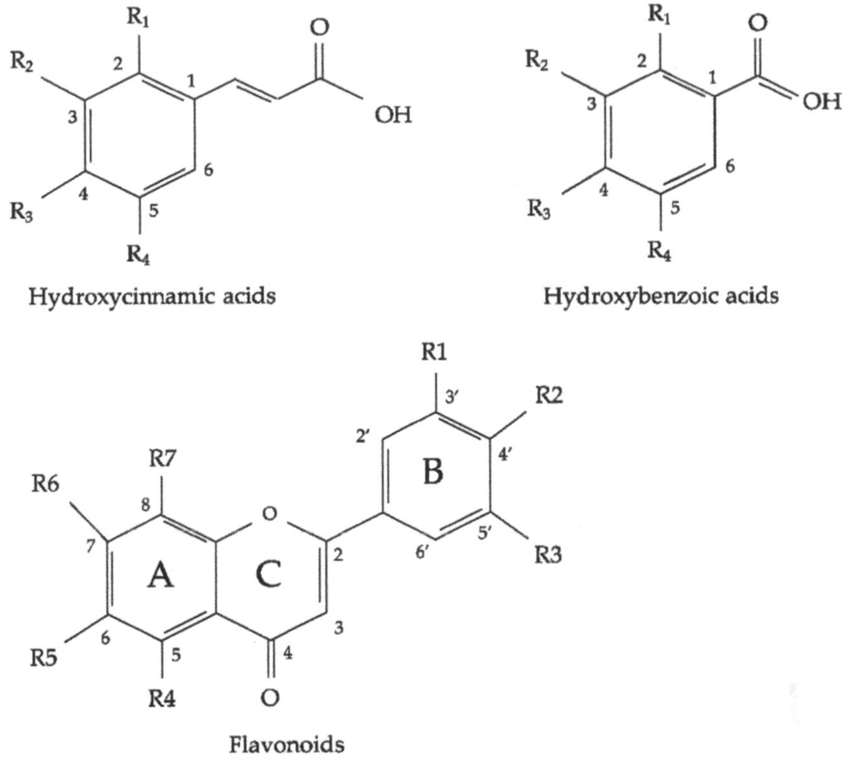

Figure 1 Basic structure of phenolic compounds.

capacity can be used as a screening tool for use in fruit breeding programmes designed to increase antioxidant phytochemicals available for human nutrition (Scalzo et al., 2005).

2.2 Anthocyanins

Anthocyanins are a group of naturally occurring pigments responsible for the red–blue colour of many fruits and vegetables. They belong to the polyphenolic flavonoids sub-class. Over 700 different anthocyanin structures have been identified so far in the plant kingdom, revealing the plasticity of the anthocyanin biosynthetic pathway; 277 of these structures have been identified in the last 20 years (Andersen and Jordheim, 2006).

Anthocyanins are particularly present in fruits (Mazza and Miniati, 1993). Temperate fruits containing anthocyanins include cherries, plums, peaches, nectarines, apples, pear, grape and berries. Anthocyanins have the structure of heterosides, the aglycon (anthocyanidin) being linked to glycosidic units. Sugars (often mono-, di- or tri-glycosides) can be, in addition, acylated with different organic acids (cinnamic, caffeic, ferulic or sinapic acid).

To date, there are 17 known anthocyanidins occurring in nature, but six principal anthocyanidins are present in most fruits and vegetables: pelargonidin, cyanidin, delphinidin, peonidin, petunidin and malvidin (Fig. 2). The glycosides of the three

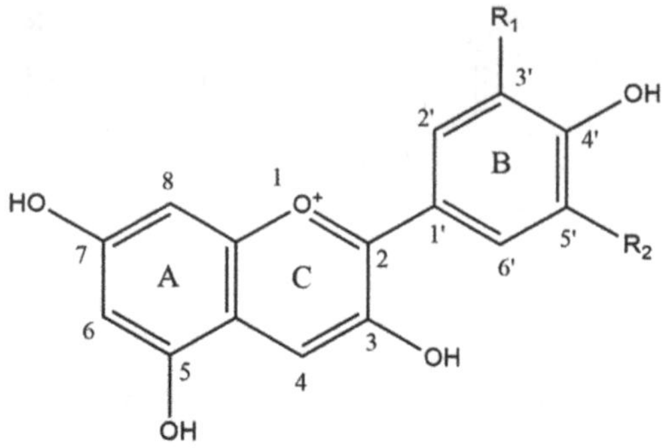

Anthocyanidins	R_1	R_2
Pelargonidin	H	H
Cyanidin	OH	H
Delphinidin	OH	OH
Peonidin	OCH_3	H
Petunidin	OH	OCH_3
Malvidin	OCH_3	OCH_3

Figure 2 Basic structure of six anthocyanidins and substitutions (R).

non-methylated anthocyanidins (pelargonidin, cyanidin and delphinidin) are the most widespread structure present in nature, while cyanidin 3-glucoside is predominant in fruits.

The hue of anthocyanins varies according to different substitution groups, and colour saturation increases with the number of hydroxyl groups and decreases with the addition of methoxyl groups. Anthocyanins are highly soluble in water, where four equilibrium forms co-exist. Under highly acidic conditions, the flavilium ion is present and relatively stable, showing a strong red colour. Under weakly acidic or neutral conditions, the red anthocyanin is converted into a highly unstable pseudo-base, the carbinol, which can be converted into the chalcone. At alkaline pH, the red flavilium cation is converted into the blue quinoidal base.

Anthocyanins can act as antioxidants thanks to their positively charged oxygen atom, which makes the structure a potent hydrogen-donating molecule, more than other flavonoids or general phenolics. Anthocyanins can act as antioxidants not only as hydrogen donors, but also as metal chelator and protein-binding compounds (Kong et al., 2003). Reduction of lipid peroxyl radicals and inhibition of lipid peroxidation have also been evaluated (Tsuda et al., 1996).

Anthocyanins have been reported to bind to DNA, forming anthocyanin-DNA complex which can protect DNA from oxidative damage after exposure to free radicals (Fig. 3) (Sarma and Sharma, 1999; Wang et al., 2011). Mas et al. (2000) suggested that anthocyanins have the ability to stabilize the DNA triple-helical structure. This feature may have *in vivo* physiological functions, and may contain the basis of the possible defence mechanism of anthocyanins against oxidative damage.

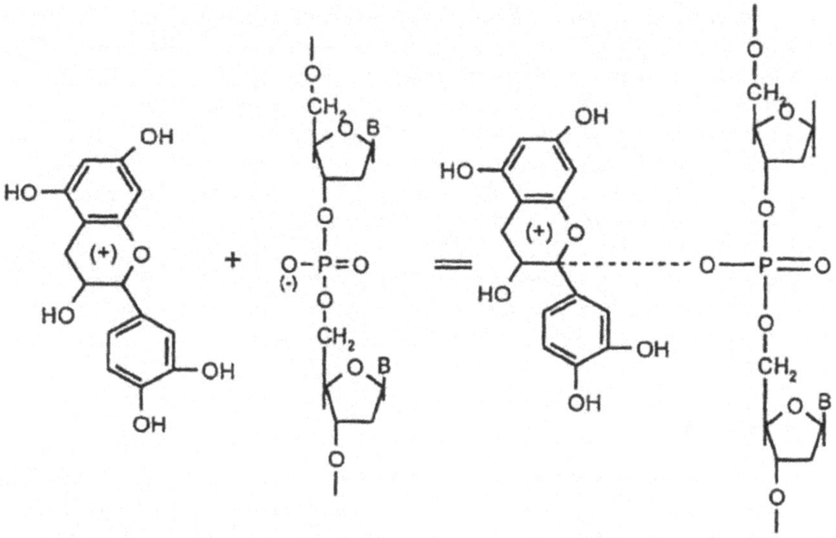

Cyanidin   DNA strand   Cyanidin-DNA complex

Figure 3 Proposed mechanism for cyanidin-DNA interaction which leads to the formation of cyanidin-DNA co-pigmentation complex. From Kong et al. (2003), by kind permission of Professor Jinming Kong.

2.3 Occurrence and role of polyphenols in fruits

Due to the increasing interest in nutraceuticals and functional foods, plant breeders have initiated selection of fruit crops with higher than normal phenolic antioxidant contents. Breeding programmes aimed at developing red-fleshed peaches [*Prunus persica* (L.) Batsch], plums (*Prunus salicina* Erhr. and hybrids) and red-peel apricots (*Prunus armeniaca* L.) with high levels of beneficial phenolic compounds for the fresh produce and processing market have been initiated in different US and EU university and research stations (Cevallos-Casals et al., 2006; Bureau et al., 2009). In pear [*Pyrus communis* L. or *Pyrus pyrifolia* (Burm.f.) Nakai], red skin colour has become a quality marker for cultivar breeding programmes, releasing new red-skinned pear varieties (Steyn et al., 2005; Feng et al., 2010 and references therein).

The different classes of phenolics which can be found mostly in fruits are discussed below.

2.3.1 Phenolic acids

Phenolic acids are the simplest phenolic compounds found in fruits. Apple (*Malus domestica* Borkh), grape (*Vitis vinifera* L.), kiwifruit (*Actinidia deliciosa* C.F. Liang & A.R. Ferguson), pears and plums are important sources of phenolic acids (hydroxycinnamic and chlorogenic acids). Red fruits are also relevant sources of gallic, p-hydroxybenzoic and protocatechuic acids, while caffeic acid is the predominant phenolic acid (75–100% of the total hydroxycinnamic acid) in most fruits (Nollet and Gutierrez-Uribe, 2018 and references therein). Phenolic monomers vary by species and account for 5–9% of the *Rubus* sp. phenolic composition consisting of phenolic acids (hydroxycinnamic and hydroxybenzoic acids), flavonol (catechin and epicatechin) and flavonol-glycosides (quercetin- and kaempferol-glycosides). Ellagic, chlorogenic and gallic acids are abundant phenolic acids in berries; blueberry (*Vaccinium corymbosum* L. and *Vaccinium angustifolium* Aiton) contains up to 2 g/kg (fresh weight) of chlorogenic acid, whereas ellagic acid accounts for about 50% of total phenolic compounds in cranberries (*Vaccinium macrocarpon* Ait.) and raspberries (*Rubus idaeus* L.) (Olas, 2018 and references therein). Generally speaking, the phenolic concentration depends on fruit types; for example, chokeberry (*Aronia melanocarpa* Michx.) is high in phenolics (2080 mg/100 g fruits) compared to blueberries and blackberries (*Rubus* sp.) (525 and 248 mg/100 g fruits), respectively (Olas, 2018 and references therein).

2.3.2 Flavonoids

Flavonoids (anthocyanins, flavonols, flavanols, flavones and isoflavone) are a large category of phenolic compounds, highly present in fruits. Anthocyanins are particularly abundant in red fruits: blackberry, blueberry, chokeberry, strawberry (*Fragaria* × *ananassa* Duch.), cherry (*Prunus avium* L. and *Prunus cerasus* L.), red grape, plum and blood orange cultivars (cvs), but are also present in some peach and nectarine cvs, some apricot cvs of recent constitution, some apple and pear cvs (Hakkinen et al., 1999; Josuttis et al., 2012; Gao and Mazza, 1995; Cantos et al., 2002; Zhu et al., 2012; Usenik et al., 2009; Lo Piero, 2015; Tomás-Barberán et al., 2001; Ruiz et al., 2005b; Mazza and Velioglu, 1992; Dussi et al., 1995). Anthocyanin accumulation during development differs among domesticated cultivars of apple, pear, peach, plum and apricot. Fruit varieties vary widely in regard to the content of phenolic (particularly anthocyanin) compounds (e.g. blond vs. blood oranges). This could be due to the genetic differences of the evaluated samples since modern fruit varieties have been obtained from breeding programmes that used parents coming from different genetic origins. As a general rule, the peel tissues contain higher amounts of phenolics, with anthocyanins and flavonols almost exclusively confined in this tissue.

2.3.3 Tannins

Ellagitannins or hydrolysable tannins are high molecular weight phenolic compounds and, as well as proanthocyanidins, are able to precipitate proteins and alkaloids. They are particularly present in red fruits such as strawberries, raspberries and blackberries. The main phenolic polymer comprises 55–71% of *Rubus* phenolic composition (Lee et al., 2012 and references therein). Proanthocyanidins or condensed tannins are oligomeric or polymeric end products of the flavonoids biosynthetic pathway and are the second most abundant natural phenolic after lignin. They are present in apples, pears, apricots, red berries, grapes, cherries and peaches (Nollet and Gutierrez-Uribe, 2018 and references therein).

Fruit phenolics are generally involved in the defence against ultraviolet radiation and pathogens (Olas, 2018). The photoprotective function of anthocyanin accumulation (induced by low temperature) in apple and pear peel has been demonstrated (Steyn et al., 2002, 2009). Resveratrol in grape is produced in response to UV radiation or microbial infection; its production decreases naturally during fruit ripening (Nollet and Gutierrez-Uribe, 2018 and references therein). This may explain the beneficial effect of using post-harvest UV-B radiation as a valid eco-friendly strategy to obtain phenolic-enriched apple fruit. UV-B treatment (219 kJ/m^2; 36 h) prior to storage (7, 14 and 21 days) increased and decreased hydroxycinnamic acids (38%) and flavonols (−45%), respectively, in apple skin. However, hydroxycinnamic acids, flavonols and anthocyanins increased in UV-B treated apples at the end of the storage period. The treatment presumably scavenged reactive oxygen species (ROS) and free radical species, thereby reducing total phenols and flavonoids (Assumpção et al., 2018). Fruit peel or skin acts as the primary defence against pathogen penetration, thereby inducing polyphenol synthesis. Thus, apple, cherry, nectarine and peach peels contain 1.6–4.5-fold more phenolic compounds than the pulp, while mango and blueberry present even higher peel/pulp ratios (32.5 and 14 folds, respectively) (Nollet and Gutierrez-Uribe, 2018 and references therein). Total phenolics and individual phenolics such as hydroxycinnamic acids (caffeic, chlorogenic and 4-*p*-coumaroylquinic acids) and flavonols (quercetin 3-glycosides) were higher in apple peel than in pulp (Jakopic et al., 2012).

Protection against (1) early feeding on fruits and (2) fungal pathogens in developing fruit tissues is provided by the presence and abundance of tannins in immature fruits. Furthermore, tannin concentration decreases during fruit ripening since the *Fahyprp* (hybrid proline-rich protein) gene related to the ripening process may be involved in polyphenols (lignin and condensed tannin) anchoring to the cell membranes (Blanco-Portales et al., 2004). Phenolic acid content has been proposed as a differentiating parameter for scab (*Venturia inaequalis*) resistance in apple cultivars due to their inherent variation observed in moderately and highly susceptible cultivars. 'Golden Delicious' (rich in phenolic acids) had less scab incidence than the highly susceptible 'Red Delicious' (90–95% disease intensity) with its low phenolic acids (tannic, caffeic, ferulic, benzoic and salicylic acids) (Singh et al., 2015). Furthermore, scab-infected 'Braeburn' apple peel contained almost 4-fold more flavan-3-ols than healthy peel (Slatnar et al., 2010).

3 Carotenoids

3.1 Carotenoid chemistry and biosynthesis

Carotenoids are the most widespread class of isoprenoid pigments synthesized by plants, algae and photosynthetic bacteria, and are responsible for the yellow, orange and red colours in various fruits and vegetables (Namitha and Negi, 2010). Carotenoids include about 600 isoprenoid compounds containing up to 15 conjugated double bonds divided into the two major groups of carotenes and xanthophylls. Carotenes, such as α-carotene, β-carotene and lycopene, are hydrocarbons that are either linear or cyclized at one or both ends of the molecule. Xanthophylls, such as β-cryptoxanthin, lutein and zeaxanthin, are the oxygenated derivatives of carotenes (Rodriguez-Amaya, 2001). The presence of a polar group in the structure (epoxy, hydroxyl and keto) alters the polarity of carotenoids and affects biological functions (Britton, 2008).

The carotenoid biosynthesis pathway is well established. The first committed step is the head-to-head condensation of two geranylgeranyl pyrophosphate (GGPP) molecules to form colourless phytoene catalysed by phytoene synthase (PSY). This is a major rate-limiting step of carotenoid biosynthesis (Cazzonelli and Pogson, 2010). The red carotenoid lycopene is synthesized by the action of phytoene desaturase (PDS), ζ-carotene isomerase (ZISO), ζ-carotene desaturase (ZDS) and carotene isomerase (CRTISO) (Naik et al., 2003). Lycopene is subsequently cyclized to synthesize α-carotene and ß-carotene, and xanthophylls are formed after a series of oxygenation reactions. After lycopene synthesis, the pathway of carotenoid biosynthesis has two main branches. In one branch, lycopene ß-cyclase (LCYb) introduces two ß-rings to lycopene to form ß-carotene. In the other branch, lycopene ε-cyclase (LCYe) and LCYb introduce one ß- and one ε-ionone ring into lycopene to form α-carotene. α-Carotene conversion to lutein occurs by sequential hydroxylation catalysed by ε-ring hydroxylase (HYe) and ß-ring hydroxylase (HYb). ß-Carotene is converted to zeaxanthin via ß-cryptoxanthin by two hydroxylation catalysed by Hyb. Zeaxanthin is converted to violaxanthin *via* antheraxanthin by zeaxanthin epoxidase (ZEP). Violaxanthin is converted into neoxanthin by the action of neoxanthin synthase (NSY).

Carotenoids are fat-soluble micronutrients that cannot be synthesized within the body and must be supplemented through daily intake to exert beneficial effects on human health and well-being. About 50 carotenoid structures are present in a typical human diet, but only 20 of them have been identified in human blood and tissues; among these, the most represented are β-carotene and α-carotene, lycopene and cryptoxanthin (Milani et al., 2017).

Carotenoids are well-known for their health benefits because of their pro-vitamin A activity. Of all the compounds, α-carotene, β-carotene and β-cryptoxanthin are the main precursors of vitamin A synthesis (Haskell, 2013). Apart from this, carotenoids are also important for antioxidant activity, intercellular communication and immune system activity (Skibsted, 2012; Stephensen, 2013). Studies indicate that consumption of carotenoid-rich fruits and vegetables may help prevent lung, head, neck and prostate cancer, cardiovascular diseases, age-related macular degeneration and cataract formation (Reboul et al., 2006; Meyers et al., 2014; Sharoni et al., 2012). Recently, lycopene- and carotene-rich oleoresins were reported to trigger significant delay of Caco-2, SAOs and NSCLC A549 malignant human cell growth (Bruno et al., 2017; Russo et al., 2017).

3.2 Occurrence and role of carotenoids in fruits

Content and types of plant carotenoids depend on several factors such as genotype, degree of maturation, cultivation method and climatic conditions. Carotenoids contribute to fruit exocarp (peel) and mesocarp (flesh) colouration. The exocarp is generally the region of highest carotenoids concentration and this is marked in citrus and apple fruits. The qualitative and quantitative distributions of carotenoid compounds vary between peel and flesh in many fruits. Mature oranges accumulate xanthophylls, which are responsible for their orange–yellow colour, while mandarin fruit are characterized by ß-cryptoxanthin and ß-citraurin that contribute to their intense orange colour. Mature white grapefruits accumulate low amount of carotenoids, mainly phytoene and violaxanthin, while the distinctive colour of red grapefruits is due to lycopene, an unusual carotene in citrus (Alquézar et al., 2013). Furthermore, in the red grapefruits, the carotenoid amount is similar in the peel and the flesh (Rodrigo et al., 2013; Alquézar et al., 2013).

Ampomah-Dwamena et al. (2012) observed that the carotenoid content was higher in peel than in flesh of commercial apple cultivars such as Royal Gala and Granny Smith, whereas in Aotea, a non-commercial cultivar, the carotenoids content was similar in the peel and flesh of the fruit. The study showed that yellow-fleshed apple genotypes accumulated more carotenoids than the white-fleshed genotypes. Carotenoid composition also differed in yellow-, red- and green-skinned commercial apple varieties (Delgado-Pelayo et al., 2014). In that study, 'Granny Smith' and 'Fuji' from France had the highest and lowest carotenoid content in the peel (151.7 and 17.1 µg/g dry weight, DW, respectively). In most apple cultivars, a characteristic carotenoid profile was observed in the peel, where lutein was the most abundant compound followed by violaxanthin, neoxanthin and ß-carotene. Nevertheless, apple flesh was characterized by high ß-carotene followed by neoxanthin and violaxanthin values. In another study, ß-carotene was the most representative compound in apples from Portugal (Dias et al., 2009), with carotenoid levels varying with species, varieties, harvest time and production site. A wide difference was reported in ß-carotene content among apple fruits of different cultivars (Tuscan, Royal Gala and Fuji Beni Shogun) and strains of Fuji (Fuji Yanfu 3, Fuji Yanfu 6 and Fuji Beni Shogun) (Jia et al., 2011). ß-carotene and total carotenoid content of the peel was higher than that of the flesh, and Fuji Beni Shogun had the highest pigment content in both the peel and the flesh. Thus, it is possible to produce higher ß-carotene content apples by selecting genotype (Jia et al., 2011).

Apricot peel generally contains higher carotenoid content than the flesh. Fruit flesh hue angle correlated with fruit flesh colour and carotenoid content of 37 apricot varieties from different genetic origins (Ruiz et al., 2005a). In that study, the total carotenoid content varied from 13.6 µg/g fresh weight (FW) in the flesh of a white variety to 38.5 µg/g FW in the peel of an orange flesh apricot variety. Furthermore, ß-carotene was the major carotenoid in apricots, followed by ß-cryptoxanthin and γ-carotene. The ß-carotene ranged from 48% to 88% in a white and orange flesh apricot. ß-cryptoxanthin ranged from 5% in an orange flesh apricot to 28% in a yellow flesh apricot. γ-carotene varied from 5% in an orange variety to 26% in a white flesh one. High levels of ß-carotene were reported in apricot by Sass-Kiss et al. (2005) and Leong and Oey (2012).

Sweet cherry fruit carotenoids consist of ß-carotene, ß-cryptoxanthin, lycopene, lutein, zeaxanthin and α-carotene (Leong and Oey, 2012; Demir, 2013). Di Matteo et al. (2016) evaluated bioactive compounds such as carotenoids from 25 autochthonous sweet cherry cultivars, selecting Mulignana and Pagliarella cultivars with a higher concentrations of carotenoids (374 and 400 µg/100 g FW, respectively).

The main carotenoids identified in peaches and nectarines were ß-carotene and ß-cryptoxanthin, with higher ß-carotene abundance than ß-cryptoxanthin, and higher total carotenoid compound in the peel of fruits compared to the flesh. As expected, yellow cultivars contained more carotenoids than white flesh cultivars; carotenoids content in yellow- and white-fleshed nectarines were 6–8 and 5–14 µg/100 g FW, respectively, and 7–20 and 71–210 µg/100 g FW in yellow- and white-fleshed peaches, respectively (Gill et al., 2002).

The carotenoid profiles of various fruit cultivars have generally been characterized to provide useful information in selective breeding programmes directed to improve their nutritional value, as well as to increase our knowledge of the carotenoids biosynthesis and accumulation in fruits. Generally, the total carotenoid content increases during ripening: carotenogenesis will take over while the chlorophylls commonly are degraded, thereby a greater amount of carotenoid compounds are synthesized in the chromoplasts than the chloroplast (Gross, 1991). Many studies have focused on carotenogenesis during development and/or amongst cultivars. Ampomah-Dwamena et al. (2012) observed changes of carotenoids accumulation pattern in different apple genotypes during fruit development. Metabolic and gene expression analysis of the carotenoid pathway suggested that expression of several rate-limiting carotenogenic steps regulate carotenoid accumulation during fruit ripening. These studies also indicated that transcriptional regulation of the phytoene gene appears to be the major regulatory mechanism responsible for determining the carotenoid pools in chromoplasts.

Carotenoid concentration and composition in citrus fruit vary greatly during maturation, and many studies have been carried out on the carotenoids biosynthesis pathway to understand regulation of colour and carotenoid composition (Kato et al., 2004; Kato, 2012; Rodrigo et al., 2013). ß-cryptoxanthin content is low in unripe citrus and it increases progressively during ripening (Kim et al., 2001). Similarly, the amount of total carotenoids and ß-cryptoxanthin was positively associated during ripening (Yoo and Moon, 2016), indicating that ß-cryptoxanthin was the major carotenoid. The content of ß-cryptoxanthin, zeaxanthin and ß-carotene increased with maturation in three different citrus cultivars grown in Korea (Yoo and Moon, 2016), while the amount of canthaxanthin, astaxanthin, lycopene and α-carotene decreased with maturation. During fruit development, total carotenoid increased three-fold in the flavedo and 10-fold in the juice sac.

The increase in carotenoid content correlates with changes in the expression of genes of the total carotenoid biosynthetic pathway (Kato et al., 2004; Rodrigo et al., 2004). The flavedo colour changed from green to orange during maturation. In the green stage, LCYe gene is highly expressed, resulting in the predominant accumulation of α-carotene and lutein (ß, ε carotenoids). During the transition of peel colour from green to orange, expression simultaneously increased in genes participating in the synthesis of ß-cryptoxanthin, zeaxanthin, all-trans violaxanthin and 9-cis violaxanthin (ß,ß-xanthophylls). In the juice sacs of the green fruit, pathway change was earlier than in the flavedo because LCYe transcripts were undetected. As fruit maturation progressed, a simultaneous increase in the expression of PSY, PDS, ZDS, LCYb and ZEP led to massive ß,ß-xanthophyll accumulation in both the flavedo and juice sacs. The results are in agreement with those reported by Fanciullino et al. (2008) for Shamouti (normal orange colour) and Sanguinelli ('blood cultivar' purple colour) cultivars, that are characterized by the highest percentages of ß,ß-xanthophylls. Thus, the carotenoid accumulation during citrus fruit maturation is due to changes in regulating carotenoid biosynthetic genes at the transcriptional level. In the flavedo of Navel oranges during fruit development and maturation, the shift from ß, ε carotenoids to the ß,ß xanthophylls was explained by the downregulation of LCYe and upregulation of LCYb and HYb (Rodrigo et al., 2004). However, Marty et al. (2005) suggested post-transcriptional regulation, end products feedback and ethylene regulation as carotenoid accumulation mechanisms in apricot fruit. Therefore, the identification of the rate-limiting steps of carotenogenesis has paved the way for breeding programmes aimed at increasing carotenoids concentration.

4 Vitamin C

4.1 Vitamin C chemistry and biosynthesis

Vitamin C is ubiquitous in all plants, but is not synthesized in humans due to gene mutations for gulonolactone oxidase, the terminal enzyme of the ascorbic acid synthesis pathway. Chemically, it is a reducing agent, donating electrons that account for all of its known physiological effects. Electrons from ascorbate can reduce metals (copper and iron) resulting in superoxide and hydrogen peroxide formation, subsequently generating reactive oxidant species (Padayatty and Levine, 2016). Vitamin C is chemically the simplest and, as such, the most industrially produced of all the vitamins. Despite being a simple molecule, its enediol structure provides it with a highly complex chemistry. For example, its redox chemistry involves comparatively stable radical intermediates that can be modified by the acidic properties of the molecule. L-ascorbic acid can act as both a redox and a complexing agent. It is the major precursor of tartaric acid in grapes as well as oxalic acid in rhubarb (*Rheum rhabarbarum*) (Davies et al., 1991).

Vitamin C is an essential, water-soluble micronutrient that exists predominantly as the ascorbate anion under physiological pH conditions. The average ascorbate concentration in the plasma of healthy humans is ~50 µM and concentrations below 11.4 µM are considered deficient and are associated with a risk of developing scurvy. The current recommended dietary allowances are 90 mg and 75 mg ascorbate for adult males and females, respectively; the daily adequate intake for infants is ~40–50 mg. Systemic ascorbate deficiency is caused predominantly by inadequate dietary intake. Over 7% of the US population is estimated to be deficient (plasma concentration < 11.4 µM) in ascorbate, according to a 2003–2004 survey (Young et al., 2015 and references therein). Vitamin C, similar to vitamin A, is considered to be a shortfall nutrient, that is defined as being under-consumed relative to the estimated average requirement or adequate intake through biochemical nutritional status indicators. These nutrient inadequacies may be associated directly with a specific health condition. In fact, ascorbate, the dominant form of vitamin C, regulates epigenomic processes demethylating the genome, presumably by mediating the interface between the genome and environment, highlighting its potential roles in various diseases (Young et al., 2015). Therefore, consumption of vitamin A- and C-rich foods, particularly fruits and vegetables, is highly recommended.

Vitamin C is almost exclusively (>90%) supplied by fruits and vegetables in the human diet. Ascorbic acid (AsA) is the predominant biologically active form of, and the main contributor to, the total vitamin C content (>85%), except for nopal (*Opuntia ficus-indica*) and avocado (*Persea americana*) (26% and 58%, respectively) (Corral-Aguayo et al., 2008). The oxidized L-dehydroascorbic acid (DHA) form contributes less than 10% of the total vitamin C (AsA + DHA) in most horticultural crops and generally increases during storage.

In higher plants, ascorbate biosynthesis and metabolism is complex and affected by several factors; changes in ascorbate status alter the signalling pathway and plant response to various stresses. AsA accumulation is not dependent on a single master regulator, which increases the challenge in vitamin C bio-fortification. AsA accumulates in fruits from glucose, predominantly through the L-galactose pathway mediated by two genes, *GDP-D-mannose pyrophosphorylase* (*VTC1* or *GMP*) and *GDP-D-mannose-3,5-epimerase* (*GME*) (Mellidou and Kanellis, 2017). *GMP* (coding for a limiting enzyme in the Smirnoff-Wheeler's pathway for AsA biosynthesis) expression is correlated with AsA concentrations in acerola (*Malpighia emarginata*) (one of the highest vitamin C containing fruit), but not in kiwifruit (*Actinidia deliciosa*) or blueberry. Apparently, other genes involved in AsA biosynthesis (*L-galactose-1-phosphate phosphatase*, *L-galactono-1,4-lactone dehydrogenase* and *D-galacturonic reductase*) as well as high *GMP* expression may be responsible for the elevated AsA in acerola (Badejo et al., 2008). However, reduction in vitamin C content of acerola due to environmental (location) effect and/or maturity stage has been ascribed to ascorbic acid oxidase enzyme, since its activity is higher in mature than immature fruits. Similarly, repressed ascorbate oxidase activity has been associated with elevated

AsA content in mature carambola (*Averrhoa carambola*) fruit (Zainudin et al., 2014). Interestingly, lower AsA acerola cultivars 'Florida Sweet' and 'Will#2' have been developed for the fresh market and juice processing since the wild varieties are considerably too tart due to high vitamin C levels (Delva, 2012). Although *GME* is not the rate-limiting step in AsA biosynthesis, its transcripts are correlated with AsA in apple and blueberry (Mellidou and Kanellis, 2017 and references therein). *GDP-L-galactose phosphorylase* (*GGP*) also regulates AsA biosynthesis (Smirnoff-Wheeler pathway) in citrus, blueberry, strawberry and kiwifruit, where *GGP* expression correlated significantly with changes in AsA levels between high- and low-AsA species. *GGP* acts synergistically with *GME*, and both genes are regulated diurnally by light (at least in kiwifruit).

In apple, grape, orange and strawberry, AsA can also accumulate through the alternate D-galacturonate reductase (GalUR) pathway (Mellidou and Kanellis, 2017 and references therein). *GalUR* expression correlates positively with AsA content in strawberry (Agius et al., 2003).

Dehydroascorbate reductase (*DHAR*) regulates AsA redox state, and its expression along with *monodehydroascorbate reductase* (*MDHAR*) correlate with AsA accumulation in blueberry. *MDHAR* silencing can increase AsA accumulation in fruits since its overexpression and silencing reduced and significantly increased AsA accumulation, respectively, in tomato (*Solanum lycopersicum*) (Gest et al., 2013). The genes/factors modulating the higher AsA levels of wild accessions, compared to modern cultivars (up to 10-fold in kiwifruit), have not been completely elucidated to date and need to be investigated. The most successful attempt to engineer elevated AsA has involved silencing of *malate dehydrogenase* (*MDH*) or overexpression of *DHAR* that elevated AsA content ~6-fold in tomato and corn (*Zea mays*), respectively (Goggin et al., 2010 and references therein).

4.2 Occurrence and role of vitamin C in fruits

Several factors influence vitamin C content in fruits including genotype, environment, cultural practices, maturity and harvesting methods, and post-harvest procedures, particularly storage conditions (Lee and Kader, 2000). AsA concentration is highly dependent on fruit species with persimmon (*Diospyros kaki*), strawberry, kiwifruit and citrus fruit considered as excellent sources. However, variations are wide within cultivars of the same species. For example, AsA content varies 29–80 mg and 14–103 mg/100 g FW in kiwifruit and berry fruits, respectively (Kanellis and Manganaris, 2014). Vitamin C contents of pulp extracts of 21 citrus fruits (orange, satsuma, clementine, mandarins, tangor, bergamot, lemon, tangelos, kumquat, calamondin and pamplemousse) commonly grown in Mauritius ranged from 166 to 677 µg/mL (Ramful et al., 2011).

Genotypic variation in vitamin C content is well documented in several fruits. Wide vitamin C range (2.5–21.4 mg/100 g FW) occurs in sour cherry that on average contains 12 mg of vitamin C/100 g FW. However, vitamin C variation was low (3–9 mg/100 g FW) in fruits of 11 Hungarian sour cherry genotypes and in three dominant, commercially grown, well-fertilized Polish sour cherry genotypes hand-picked at maturity (17.7–21.1 mg/100 g FW). Furthermore, fruits of the most and least productive Polish cultivars displayed the highest and least vitamin C contents, respectively (Borowy et al., 2018 and references therein). In kiwifruit, vitamin C content varies between and within species: 50–450 mg and 20–250 mg/100 g for *Actinidia chinensis* and *A. deliciosa*, respectively (Huang et al., 2004). However, vitamin C contents of 'SunGold' (*A. chinensis* var. *chinensis* 'Zesy003') and 'Sweet Green' (*A. chinensis* var. *chinensis* × *A. chinensis* var. *deliciosa* 'Zesh004') (161 and 150 mg/100 g edible flesh, respectively) were higher than those of the common green variety 'Hayward' (*A. deliciosa*) (85 mg/100 g) (Sivakumaran et al., 2018). Russian sea buckthorn cultivars (*Hipophea rhamnoides* subsp. *mongolica*) grown in Poland and high in AsA content (53–131 mg/100 g FW) may range from 0.5 to 3.3 g/kg, differing from those of subsp. *turkistanica* (2.5–4.2 g/kg) (Teleszko et al., 2015). Total vitamin C (AsA + DHA) content of 31 apple cultivars varied 3.6-fold (401–1448 nmol/g FW) grown at the same location in Belgium; this indicates the limited genetic potential for breeding for increased fruit L-AA contents (Davey and Keulemans, 2004). Furthermore, a genotype-environment interaction was also important in vitamin C concentration (320 vs. 75 µg/g) indicating that breeding lines should be evaluated in multiple locations and seasons (Patil et al., 2014).

AsA content varied (19.3–71.5 mg/100 g FW) in six strawberry cultivars from four locations (Lee and Kader, 2000) and was not affected by altitude when grown in Italy under the same cultural and environmental practices (Doumett et al., 2011). The climate can considerably affect AsA content of fruits. For example, the subtropical climate, along with soil composition, accounted for the higher (2- to 7-fold for strawberry and blueberries, respectively) AsA content of berries grown in Brazil compared to temperate production zones. Furthermore, blackberry (*Rubus* spp.) plants produce large quantities of fruit in subtropical areas, with some varieties producing higher amounts of vitamin C compared to those in temperate zones (De Souza et al., 2014). Similarly, native wild Chinese sea buckthorn berries (*Hipophea rhamnoides* subsp. *sinensis*) contain 5–10 times more vitamin C in the juice fraction than those of European and Russian subspecies (*rhamnoides* and *mongolica*, respectively) (Kallio et al., 2002).

Growing season also can affect AsA content. For example, Argentinian-grown oranges harvested in 2011 had significantly higher AsA content (55 vs. 49 mg/100 mL) than those harvested in the preceding year (2010) notable for its high precipitation (Stinco et al., 2015). However, ascorbate concentration in six highbush blueberry cultivars differed less than 15% between two growing seasons in Poland, except for 'Duke' with ~85% drop between seasons (Łata and Wi ska-Krysiak, 2010). Pulp vitamin C content did not differ significantly in similar citrus varieties harvested at different periods except for tangor, tangelo, kumquat and pamplemousse (Ramful et al., 2011).

Developmental stages influence AsA content such that it increases from the early stages of development to full maturity in a cultivar-dependent manner in strawberries grown in the same environment in Brazil. However, the final total AsA content (40–85 mg/100 g) at maturity was not related to pattern of accumulation in the tissues during fruit development (Cordenunsi et al., 2002). Furthermore, AsA and anthocyanin contents in strawberry are inversely associated such that these components are presumed to be complementary, probably as antioxidant agents protecting cells against oxidative damage (Cordenunsi et al., 2002). In pomegranate (*Punica granatum*), AsA decreases with fruit maturation and ripening (Erkan and Kader, 2011).

AsA is one of the major non-enzymatic antioxidants in plants, involved in both enzymatic and non-enzymatic quenching of ROS, particularly in the scavenging of hydrogen peroxide (Suza et al., 2010). Most fruits with high AsA content (strawberries, bananas [*Musa* spp.], lemons and oranges) also exhibited high antioxidant capacity (Kevers et al., 2007). It correlates highly with free radical scavenging activity of commercial orange juices and antioxidant activity (TEAC, r = 0.73) of fresh orange juice. Vitamin C in hydrophilic extracts of tropical fruits (avocado, black sapote, guava, mango, papaya, prickly pear and strawberry) correlated highly ($r \geq 0.88$) with antioxidant activity evaluated using the DPPH, DMPD, FRAP and TEAC methods (Corral-Aguayo et al., 2008). Vitamin C contributes 65–100% of antioxidant activity of beverages derived from citrus fruits (Ramful et al., 2011 and references therein) and only 0.35–8.6% for eleven fruits (cranberry, apple, red grape, strawberry, pineapple, banana, peach, lemon, orange, pear and grapefruit) (Sun et al., 2002). AsA is poorly or not correlated with antioxidant activity of many fruits (nectarines, peaches, plums) and exotic tropical fruits; it is inversely ($r = -0.80$) associated with antioxidant activity for strawberries, raspberries and blueberries (Ramful et al., 2011 and references therein).

5 Production practices that influence bioactive compound synthesis

Genetic characteristics of cultivars strongly determine their BC contents. Landraces and wild unexplored genotypes often accumulate high BCs that may be valuable for breeding purposes. However, fruits with high secondary metabolites (e.g. total phenolics) may confer bitter and astringent tastes (Manganaris et al., 2018 and references therein). Total polyphenolic content of wild apple germplasm varies considerably and is generally higher than that of commercial cultivars, similar to the trend for AsA. Old apple cultivars also contain higher polyphenolic and vitamin C contents of flesh and antioxidant capacity relative to new cultivars (Kschonsek et al., 2018). The lower phenolic and flavonol contents of new apple cultivars are presumed to result from breeding for their consumer-preferred sweet flavour and low enzymatic browning. The higher average content of total polyphenols, particularly flavonols, of old apple cultivars may also be due to species diversity (Kschonsek et al., 2018).

Breeding strategies to enhance polyphenolic content can also improve quality traits (colour, aroma, disease resistance and browning disorders) related to the phenylpropanoid pathway (Manganaris et al., 2018). Breeding strategies should focus on the greater within-family, rather than among family, variation in phenolics, including assessments of genotypic performance in several locations and years to differentiate environmental effects. The phenolic compositions among apple cultivars are compounded by differences in growing regions, highlighting that certain combinations of cultivar and growing region can result in significantly higher levels of phenolic compounds (Brovelli, 2006 and references therein). In strawberry, the role of genetic background (different cultivars) has been demonstrated in relation to their antioxidant activities (Scalzo et al., 2005).

Classical genetic breeding is time-consuming and the release of new cultivars with desired characteristic takes many years. The genetic background of the plant is the main factor determining the quality and content of phenolic compounds in plant tissues. Even if the genetic approach relies on the variability existing in the genetic pool, and therefore is more potent to create high BC content cultivars, the development of agronomic strategies to enhance the production of useful metabolites provides an opportunity to optimize the currently available cultivars. The agronomic approach can be valuable to achieve a moderate (~two-fold) increase by changing only the crop growing conditions (Poiroux-Gonord et al., 2010).

5.1 Polyphenols: environmental variation and agronomic conditions

Over the past few years, many of the genes coding for enzymes involved in the phenylpropanoid pathway have been identified, cloned and characterized. Scientists have been able to dissect the regulatory mechanisms of the phenylpropanoid pathway, increasing our understanding of the abiotic and biotic factors governing polyphenols accumulation. The information can now be applied to enhance or optimize the production of health benefitting polyphenols in plants. Apart from the transgenic approach (which is still problematic for consumer acceptance), the phenylpropanoid pathway can also be manipulated through agronomic and physiological factors involving exposure to specific light regimes, nutrient supply, microbe–host plant interaction, elicitor applications, cultural management and water stress imposition (Dorais and Ehret, 2008; Jaganath and Zainal, 2016). In the following paragraphs, we highlight the environmental factors that can contribute to the modulation of polyphenols, particularly anthocyanins, in fruits.

5.1.1 Effect of light and temperature on the polyphenols (anthocyanin) content

One of the factors that dramatically influence polyphenols (particularly anthocyanins) content is light exposure, both photoperiodic as well as quality and intensity. Fruit bagging and shading experiments have demonstrated the importance of light conditions on the biosynthesis and accumulation of anthocyanins (Zoratti et al., 2014a). The anthocyanin function in vegetative (photosynthetic) tissues has been proposed as being photoprotective, since it has been shown that anthocyanins reduce photoinhibition and photobleaching of chlorophyll under high light stress conditions (Steyn et al., 2002). Similarly, anthocyanins afford photoprotection to fruit peel (apple and pear) during low-temperature-induced light stress and the protection is not a fortuitous side effect of light absorption by anthocyanin (Steyn et al., 2009).

The regulation of flavonoid biosynthesis in fruit crops has been studied extensively and reviewed during the last 10 years, leading to knowledge of the key transcription factors that coordinate and regulate flavonoid structural genes (Jaakola, 2013). One of these transcription factors (R2R3 MYB) has been found to be inducible with specific light wavelengths, thus leading to distinct flavonoid biosynthesis in apple, pear, nectarine, Chinese bayberry (*Myrica rubra*), strawberry, litchi (*Litchi chinensis*) and grapevine (Kim, 2003; Zoratti et al., 2014a). In grape berries, the anthocyanin content was significantly enhanced by blue light filter film and suppressed by red, orange and green films compared to white film (control) (Cheng et al., 2015). High light intensity upregulates the expression of an array of both early and late flavonoid biosynthesis genes in the grape berry skin, leading to an increased content of anthocyanins and general flavonoids (Zoratti et al., 2014a). Moreover, following the illumination with blue, red, far-red or white light during the berry ripening process in bilberry, it has been shown that all the three monochromatic lights influenced the accumulation of total anthocyanins in the mature fruits (Zoratti et al., 2014b). Therefore, in controlled production conditions (e.g. greenhouses, plastic covers, high tunnels), it is crucial to use the correct spectrum of light if an increase of anthocyanin is desired (see Fig. 4).

Figure 4 'Robada' apricot grown under a 'Luminence THB' polyethylene-covered high tunnel (left) and in the open orchard (right) in Michigan, USA. Photos by G. Lang.

Flavonoid biosynthesis is influenced by light in a species- and cultivar-specific manner. It is possible that white-fleshed apples have lost the ability to produce pigmentation throughout the fruit due to a mutation in one or more biosynthetic or regulatory gene(s), as has been observed for white grapes (Boss et al., 1996). Not all species require strong light exposure to accumulate anthocyanins. Bilberry (*Vaccinium myrtillus*), one of the best sources of anthocyanins, grows and produces better in shaded light conditions. Light is recognized as an important regulation factor in numerous crops belonging to the *Rosaceae* family, including strawberry, peach, apple and pear (Esplay et al., 2007; Feng et al, 2010; Zoratti et al., 2014a).

Another environmental factor that can significantly influence anthocyanin accumulation in fruit crops is temperature. Low night temperatures induce anthocyanin synthesis in grape berries, with higher *UFGT* gene expression at low temperature (Mori et al., 2005), while higher temperatures inhibit anthocyanin accumulation (inhibiting PAL activity) or promote their degradation (De Pascual-Teresa and Sanchez-Ballesta, 2008). Studies in apple and pear support the view that anthocyanins are adaptable light screens deployed to modulate light absorption in sensitive tissues such as fruit peel in response to environmental triggers such as cold fronts (Steyn et al., 2002). Blood orange varieties (Tarocco, Moro and Sanguinello) require a wide day–night thermal range to induce the biosynthesis of anthocyanins and maximize colour formation. Production of red oranges characterized by high anthocyanin levels is limited to areas with characteristic climate conditions, such as the area around Mount Etna in Sicily (Italy), where high colour intensity and quality blood orange fruits are produced (Lo

Piero, 2015). Cooling by micro-sprinkler irrigation has been found to increase red colour and anthocyanin content in apple (Iglesias et al., 2005).

5.1.2 Effect of cultural practices on the polyphenols content

Many cultural practices influence fruit polyphenolic content. Moderate pruning, cluster thinning or regulated irrigation significantly increased phenolic content of grapes (Ruiz-García and Gómez-Plaza, 2013). The pruning presumably imparts shock/stress to the plants, thereby elevating phenolic synthesis. Furthermore, pruning can promote other desirable characteristics, such as high fruit firmness and reduced disease incidence (Asrey et al., 2013). It is well-known that environmental stress, such as drought, diverts considerable amounts of substrates from primary to secondary metabolism, leading to a reduced growth rate and competitive ability of the plant, while accumulating various phenolic compounds (Dixon and Paiva, 1995). Application of water stress during different phases of crop development must be managed in such a way as not to hinder plant development. Mild stress treatments seem to be promising for increasing BC content, without negatively affecting yield (Poiroux-Gonord et al., 2010).

Generally, phenolic compounds and anthocyanins increase with reduced water supply, enabling the use of deficit irrigation strategies to improve fruit quality and water use efficiency. In grape, water stress applied during the pre-veraison period increased the concentration of phenolic compounds in wines (Intrigliolo and Castel, 2010). Water management/stress (15% of water potential) can elevate grape skin tannins and anthocyanins, and also modify/shift grape anthocyanins composition and profiles (Tavarini et al., 2011 and references therein). Other investigations on peaches and apricots have shown that deficit irrigation during fruit growth can have a positive effect on fruit quality (Crisosto et al., 1994; Torrecillas et al., 2000; Pérez-Pastor et al., 2007). In pomegranate, deficit irrigation at specific times (fruit set or growth) enhanced fruit quality attributes, such as soluble solids and anthocyanin content. Depending on the phenological period of application of water shortage, deficit irrigation can be used as a field practice to control ripening attributes (Laribi et al., 2013). However, high phenolic levels due to water stress can result in internal browning of stone fruits. Moreover, the drought-induced variations in gene expressions for polyphenols production were highly cultivar-specific. Water stress increased total hydroxycinnamic acid of peach depending on cultivar and rootstock (Tavarini et al., 2011).

It is known that physiological responses of fruit trees are influenced by rootstock, because trees on different size-controlling rootstocks result in different water relations, gas exchange and vegetative growth. Even if rootstock has a much greater effect on tree growth and yield, it can also influence fruit quality (Castle, 1995), particularly phytochemicals such as phenolics in apricots (Bartolini et al., 2014) and antioxidant activity in peach (Scalzo et al., 2005). In cherries, the regulation of fruit quality is more dependent on the cultivar (and the crop load), although the effects of rootstock genotype and cultivar × rootstock interaction were also statistically significant (Gonçalves et al., 2006). In peach, total antioxidant capacity and levels of some phytochemicals (phenols, ascorbic acid and β-carotene) are significantly influenced by rootstock, even if it is not possible to define a common behaviour in terms of rootstock vigour. Indeed, rootstocks of similar vigour produced fruits with very different nutritional characteristics, indicating that the rootstock effect is more complex than just vigour (Remorini et al., 2008).

5.1.3 Effect of nutrient management on the polyphenols content

Nutrient management is an important feature of healthy plant production for food purposes. Optimal mineral nutrition is a prerequisite for providing co-factors for many enzymes of the flavonoid pathway. Carbohydrate availability is another prerequisite for phenylpropanoid accumulation. Reduction or accumulation of various phenolic compounds in plant tissues has been reported in relation to specific nutrients. For example, nitrogen (N) supplementation was associated, in apple, with reduction of anthocyanin and flavan-3-ol levels (Awad and Jager, 2002). Conversely, excess N may inhibit anthocyanin production in green apple cultivars such as 'Granny Smith' where red colour is undesirable. Phosphorus supplementation increased the percentage of red skin (i.e. anthocyanin content) at harvest for different apple cultivars (Paliyath et al., 2002).

Arbuscular mycorrhizal fungi (AMF) refers to symbiosis between plants and soil-borne fungi. Several examples demonstrate enhanced synthesis of phenolic compounds in mycorrhiza-inoculated plants. Higher anthocyanin content was observed in fruits from strawberry plants inoculated with *Glomus intraradices* (Castellanos-Morales et al., 2010). AMF can therefore be used as a strategy not only to improve the growth of crops, but also to enhance the production of specific phenolic compounds.

Organic fertilization can directly result in lower N content and pesticide residues, and a corresponding increase in content of phenolic defence compounds (Veberi , 2016). This can increase plant resistance to pests and diseases, although at the cost of a lower growth rate and yield (Brandt and Mølgaard, 2001). Organic fertilization increases phenolic content in apple, marionberry (blackberry), strawberry and white seedless grape compared to conventional fertilization. White seedless grapes managed under organic fertilization exhibited the highest total phenolic, catechin, quercetin-3-galactoside, total flavonoids and antioxidant (DPPH) activity compared to those with conventional or integrated fertilization. However, physical fruit characteristics of white seedless grapes under conventional fertilization were considerably higher than those with organic or integrated fertilization (Koureh et al., 2018).

Apples in organic production contained higher concentrations of hydroxycinnamic acids, flavan-3-ols, flavonols and total phenolics, particularly in the peel, than those in integrated or improved integrated pest management (IPM) production systems (Jakopic et al., 2012). Organically grown strawberries ('Candonga' and 'Ventana' genotypes) exhibited significantly higher antioxidant activity than IPM strawberries (Blando et al., 2012). The higher phenols concentration in organically grown fruits can be attributed to severe stress (biotic or abiotic) that induces phenolic biosynthesis. However, other factors, such as ultraviolet radiation, low temperature, nutrient deficiencies and pest or pathogen attacks, may contribute to differences in concentration of fruit phenolic compounds independent of production system/management.

5.1.4 Effect of treatments with elicitors, stimulating agents and plant activators on polyphenols content

Physical elicitors (e.g. high/low temperatures, light (UV, gamma radiation)) can trigger plant defence mechanisms that may concomitantly increase polyphenolic contents of fruit, such as grapes and blueberries (Ruiz-García and Gómez-Plaza, 2013). Phenolic acid contents of cherries generally decrease during storage at 1–2°C and increase at 15 ± 5°C, while anthocyanin levels increase at both temperature ranges without affecting flavonols and flavan-3-ol contents (Valero, 2015). Total phenolics and phenolic compounds (neo-chlorogenic, hydroxycinnamic and 3-p-coumaroylquinic acids) increase with cherry maturity and during storage.

Chemical elicitors (chitosan, benzothiadazole [BTH], harpin and 1-methylcyclopropane) that stimulate salicylic acid (SA) and jasmonic acid (JA) signalling can activate plant defence responses and PAL, thereby elevating phenolic content in various fruits [apples, bananas, grapes, mangoes, peaches and berries (strawberries, raspberries, blackberries, bayberries)]. SA activates the phenylpropanoid pathway, while JA induces de novo synthesis of phenolic compounds; both confer resistance to plant diseases (Ruiz-García and Gómez-Plaza, 2013). Other elicitors include oxalic acid (OA), calcium chloride (enhances the defence-related enzymes-β-1–3-glucanase and PAL) and potassium silicate. SA (or its analog acetylsalicylic acid) increases total phenolic acid content in fruits including sweet cherry, pomegranate and plum (Valero, 2015).

Pre-harvest applications (foliar sprays) of SA (0.5 mM) or acetyl salicylic acid (1 mM) to sweet cherry trees elevated the total phenolics, total anthocyanins and hydrophilic total antioxidant activity of fruit at harvest (commercial ripening stage) and during post-harvest storage (Giménez et al., 2017). Furthermore, the treatments also enhanced the activity of antioxidant enzymes (catalase, peroxidase, superoxide dismutase and ascorbate peroxidase), thereby delaying the post-harvest ripening process. OA increases defence-related secondary metabolites, such as phenolics, and effectively delays fruit ripening in plums and peaches by inhibiting ethylene biosynthesis. Pre-harvest OA treatment (2 mM) of cherry trees increased total phenolics and anthocyanin concentrations considerably during the last 2 weeks of on-tree fruit ripening. Similar increases in total phenolics have been reported as a result of post-harvest treatment during cold storage of sweet cherry. Post-harvest treatment (1 mM SA, ASA or OA) before cherry fruit storage can delay the post-harvest ripening process and, additionally, maintain higher levels of antioxidants (phenolics and anthocyanins) (Valero, 2015). Pre-harvest BHT treatment increased anthocyanin and other flavonoid biosynthesis (resveratrol) in grape (Iriti et al., 2004).

Plant hormones, such as abscisic acid (ABA), seem to affect phenylpropanoid biosynthesis. ABA applications on grape and cherry increased the anthocyanin content of the skin (Kondo and Inoue, 1997). The role of ethylene in anthocyanin biosynthesis has not been fully elucidated: an ethylene-related increase of PAL activity was described in the first studies around 1970–80; recently ethylene-independent anthocyanin biosynthesis has been proposed in apples and cherries (Treutter, 2010 and references therein).

Substances generally known as 'biostimulants' have recently been considered to activate secondary metabolite biosynthesis in plants. A biostimulant is any substance and/or microorganism that influences plant physiology and provides a health benefit for growth, development, response to abiotic stress and improvement of quality (Calvo et al., 2014; Du Jardin, 2015). For example, a commercial new chitosan formulation not only protected grapevines from powdery mildew (Uncinula necator Schwein.), but also improved total phenols and antioxidant activity in both fruits and wine (Iriti et al., 2011). A commercial oak aqueous extract has been used on grapes at veraison, affecting fruit composition and resulting in higher polyphenols content of the wine, which was less alcoholic and acidic with high colour stability (Pardo-García et al., 2014). Oak aqueous extracts foliar application to the grapevine (Martínez-Gil et al., 2011) and more recently vine-shoot extracts (Sánchez-Gómez et al., 2017) have been used as 'biostimulants' to improve grape yield and wine quality, particularly in aromatic cultivars.

5.2 Carotenoids: effects of light, sugars and plant probiotic bacteria on carotenoid accumulation

Light and sugars are involved in fruit ripening, and can affect carotenoid accumulation in fruits such as citrus, strawberry and grape. In citrus species, light exposure during fruit ripening generally enhances carotenoid accumulation and peel colour (Cronje et al., 2011, 2013). In contrast, Lado et al. (2015) observed that red grapefruit developed more intense red colouration when grown under low light conditions. Zhang et al. (2012) showed that blue, but not red, light induced total carotenoid accumulation in Satsuma mandarin, Valencia orange and Lisbon lemon. Furthermore, *PSY* expression was upregulated by blue light treatment. The results showed that *PSY* and *LCYb* gene expression was increased, respectively, by sucrose and mannitol treatment, contributing to increased carotenoid content. Carotenoid metabolism during bilberry fruit development under different light conditions showed a metabolic flux towards formation of β-branch of carotenoid compounds, along with increased enzymatic carotenoid cleavage and degradation (Karppinen et al., 2016).

Recently, Rahman et al. (2018) evaluated the effects of plant probiotic bacterial strains BChi1 and BRRh-4 on various types of bioactive compounds such as carotenoids in fresh strawberry fruits. Application of plant probiotic bacteria increased total carotenoid content compared to the non-treated control. The highest carotenoid content, expressed as mg of lutein equivalent, was recorded in plants treated with BRRh-4 (7.71 mg lutein/g FW), followed by treatment with BChi1 (6.46 mg lutein/g FW) and the control plants (2.82 mg lutein/g FW).

5.3 Vitamin C: environmental variation and agronomic conditions

Plants can be considered as 'living bioreactors' for vitamin C production. Moreover, the baseline production can be increased as a response to stress conditions (Locato et al., 2013). During the growing season, light intensity quantitatively influences AsA accumulation; low light intensity exposure results in low-AsA content in plant tissues (Lee and Kader, 2000). Therefore, fruit on the same plant exposed to maximum sunlight contains higher vitamin C levels than those in shade or the interior of the canopy. This may account for the beneficial effects of tree pruning and thinning on vitamin C content of fruits. Other agronomic practices that enhance light interception, such as highly reflective mulches, significantly increase total vitamin C levels of strawberries (Atkinson et al., 2006); for example, white (Extenday) reflective mulch increased fruit AsA concentration in 'Elsanta' by 11% and 14% over hessian and black polythene, respectively. Branch girdling, another extensively used practice to promote flowering and improve fruit set in citrus production, significantly increased AsA content in Satsuma mandarin fruit peel or pulp (Yang et al., 2013). This elevated AsA accumulation is presumably due to sugar or phytohormone modulation that regulates AsA biosynthesis-related gene expression in fruit tissues.

Temperature also influences AsA content: mandarins grown under cool temperatures (20–22°C day, 11–13°C night) contain higher vitamin C than at hot temperatures (30–35°C day, 20–25°C night). Similarly, grapefruits grown in coastal areas of California generally contain more vitamin C than those grown in the desert (Lee and Kader, 2000). AsA content generally decreases gradually as storage temperature or duration increases; in apples, AsA content can drop to less than 50% of the original amount during cold storage after

6 months (Lee and Kader, 2000 and references therein). In most fruits, AsA content is relatively stable but decreases rapidly during the first day of storage, for example apricot, banana, melon, cherry and citrus fruits (Kevers et al., 2007). Cold storage (6°C, 65 days) positively influenced vitamin C content in blond orange varieties, but reduced its content in blood oranges (Rapisarda et al., 2008). Longer cold storage (1°C, 3 months) reduced total AsA content (35%) of apples, but it increased slightly in some cultivars, suggesting an acclimation response to low temperature (and other stress) (Davey and Keulemans, 2004). Strawberries lost 20–30% of total AsA during storage (1°C, 8 days) (Lee and Kader, 2000 and references therein). AsA loss occurs during cold storage in chilling-sensitive fruits such as pomegranate; however, AsA increases significantly (11.5%) following OA (6 mM) application to cold stored (2°C, 84 days) pomegranates (Sayyari et al., 2010).

Environmentally sustainable agricultural practices, such as organic agro-ecosystem production, elevates vitamin C (AsA + DHA) and AsA content (9.7%; 0.621 vs. 0.566 mg/g FW; p = 0.009), and AsA/DHA ratio (>3-fold) compared to conventionally grown strawberries in California (Reganold et al., 2010). This favourable antioxidant status of the organically grown strawberries increases its anti-proliferative effects on breast (MCF-7) and colon (HT-29) cancer cells compared to conventionally grown berries, with AsA concentrations inversely correlated with cancer cell proliferation (Olsson et al., 2006). Other pre-harvest practices aimed at delaying fruit ripening, such as foliar K treatments, also increase AsA content (21%) in muskmelons grown in high K soils (Jifon and Lester, 2009). The positive response of AsA to foliar K applications is presumably due to improved metabolic processes such as leaf photosynthesis, sugar production, photoassimilate transport from leaves to fruits and substrate availability for AsA biosynthesis. Calcium treatments (2% and 4% $CaCl_2$) increased AsA content (up to 29% and 68%, respectively) in apples as well as increased fruit firmness (Lee and Kader, 2000 and references therein) and in pomegranate (Erkan and Kader, 2011). Post-harvest application of SA to delay ripening and maintain fruit quality attributes generally increases antioxidant enzymes during storage. The changes in AsA content in response to SA treatment differ depending on the fruit; its level is maintained in pomegranate exposed to SA (Sayyari et al., 2010). Similarly, JA application elevated AsA levels (140%) in Japanese plum (Khan and Singh, 2007).

6 Future trends and conclusion

It is important to underline the economic impact of managing/elevating the content and profile of BCs in fruit crops and their related health benefits. Consumers' choices and market trends can follow different paths. Fruit colour often is a consumer selection criterion, for example selection of red apple strains in spite of a higher price, even if red colour is not always associated with high fruit quality (Treutter, 2010).

Research progress, through genetics and agronomic practices, can influence and enhance the content of BCs in fruits. 'Classical' genetic breeding is time-consuming and the release of new cultivars with desired BC characteristics takes years. Genetic engineering and modification of genetic traits (GMO) has not been successful, at least in Europe. Someday, 'genome editing' and subsequent 'cis-genesis' might be more readily accepted, both by consumers and governmental agencies.

Plant physiological research can improve agronomic strategies to increase BC contents. Studies on the adaptation and response of physiological processes are fundamental in designing crop production techniques that can yield reliable and realistic increases in BC contents in fruits. From basic research, new mathematical models fitting the interaction between environment × physiological processes for a specific organ (fruit) and genome (cv) need to be developed. New production technologies, such as biostimulants tailored to a specific target, are expected to emerge, as well as new post-harvest and storage solutions which enable preservation and/or improvement of fruit quality in terms of BC contents.

From a grower's point of view, it is important to assess the feasibility of the agronomic approach (at both the technical and economic levels). In fact, the increase of BC contents, for example by means of reduced N application or moderate drought stress, can incur yield reductions. The balance between the higher price paid for a higher quality fruit and a lower yield needs to be always positive.

7 Where to look for further information

In this section readers can find references helping them to explore further on bioactive compounds in fruits and their sustainable elicitation.

Books:

- Mazza, G. and Miniati, E. (Eds) (1993), *Anthocyanins in Fruits, Vegetables and Grains*, Boca Raton, FL, USA: CRC Press.
- Yahia, E. M. (Ed.) (2017), *Fruit and Vegetable Phytochemicals: Chemistry and Human Health*, 2nd Edition, John Wiley & Sons, Ltd. doi:10.1002/9781119158042.

Reviews:

- Gonzalez-Aguilar, G., Robles-Sánchez, R., Martínez-Téllez, M., Olivas, G., Alvarez-Parrilla, E. and De La Rosa, L. (2008), 'Bioactive compounds in fruits: Health benefits and effect of storage conditions', *Stewart Postharvest Review*, 4(3), 1–10. doi:10.2212/spr.2008.3.8.
- Gniech Karasawa, M. M. and Mohan, C. (2018), 'Fruits as prospective reserves of bioactive compounds: A review', *Natural Products and Bioprospecting*, 8, 335–46. doi:10.1007/s13659-018-0186-6.

General information on bioactive compounds and functional foods can be found in different useful websites, referring to research organizations:

- http://www.choosecherries.com/cherry-marketing-institute/
- https://www.functionalfoodscenter.net/
- http://www.inaf.ulaval.ca/

Below are some references to universities or research centres where international research projects on bioactive compounds are carried out:

- Michigan State University, Department of Horticulture, East Lansing, MI, USA
- Université Laval, Quebec, Canada; University of British Columbia, Faculty of Land and Food Systems, Vancouver, Canada
- University of Helsinki, Department of Food and Nutrition, Helsinki, Finland; Natural Resources Institute Finland (LUKE), Helsinki, Finland

These are only few examples; it must be emphasized that many universities, all over the world, deal with programmes on nutraceuticals.

8 References

Agius, F., González-Lamothe, R., Caballero, J. L., Muñoz-Blanco, J., Botella, M. A. and Valpuesta, V. (2003), 'Engineering increased vitamin C levels in plants by overexpression of a D-galacturonic acid reductase', *Nat. Biotechnol.*, 21, 177–81. doi:10.1038/nbt777.

Alquézar, B., Rodrigo, M. J., Lado, J. and Zacarías, L. (2013), 'A comparative physiological and transcriptional study of carotenoid biosynthesis in white and red grapefruit (*Citrus paradisi* Macf.)', *Tree Genet. Genom.*, 9, 1257–69.

Ampomah-Dwamena, C., Dejnoprat, S., Lewis, D., Sutherland, P., Volz, R. K. and Allan, A. C. (2012), 'Metabolic and gene expression analysis of apple (*Malus × domestica*) carotenogenesis', *J. Exp. Bot.*, 3, 4497–511. doi:10.1093/jxb/ers134.

Andersen, Ø. M. and Jordheim, M. (2006), 'The anthocyanins', *in* Andersen, Ø. M. and Markham, K. R. (Eds), *Flavonoids: Chemistry, Biochemistry and Applications*, Boca Raton, FL: CRC Press, pp. 471–551.

Asrey, R., Patel, V. B., Barman, K. and Pal, R. K. (2013), 'Pruning affects fruit yield and postharvest quality in mango (*Mangifera indica* L.) cv. Amrapali', *Fruits*, 68, 367–80.

Assumpção, C. F., Hermes, V. S., Pagno, C., Castagna, A., Mannucci, A., Sgherri, C., Pinzino, C., Ranieri, A., Flôres, S. H. and De Oliveira Rios, A. (2018), 'Phenolic enrichment in apple skin following post-harvest fruit UV-B treatment', *Postharvest Biol. Technol.*, 138, 37–45.

Atkinson, C. J., Dodds, P. A. A., Ford, Y. Y., Le Mière, J., Taylor, J. M., Blake, P. S. and Paul, N. (2006), 'Effects of cultivar, fruit number and reflected photosynthetically active radiation on *Fragaria × ananassa* productivity and fruit ellagic acid and ascorbic acid concentrations', *Ann. Bot.*, 97, 429–41.

Awad, M. and Jager, A. (2002), 'Relationships between fruit nutrients and concentrations of flavonoids and chlorogenic acid in "Elstar" apple skin', *Sci. Hort.*, 92, 265–76.

Badejo, A. A., Tanaka, N. and Esaka, M. (2008), 'Analysis of GDP-D-mannose pyrophosphorylase gene promoter from acerola (*Malpighia glabra*) and increase in ascorbate content of transgenic tobacco expressing the acerola gene', *Plant Cell Physiol.*, 49, 126–32.

Bartolini, S., Leccese, A., Iacona, C., Andreini, L. and Viti, R. (2014), 'Influence of rootstock on fruit entity, quality and antioxidant properties of fresh apricots (cv Pisana)', *New Zeal. J. Crop Hort. Sci.*, 42(4), 265–74. doi:10.1080/01140671.2014.894919.

Blanco-Portales, R., López-Raéz, J. A., Bellido, M. L., Moyano, E., Dorado, G., González-Reyes, J. A., Caballero, J. L. and Muñoz-Blanco, J. (2004), 'A strawberry fruit-specific and ripening-related gene codes for a HyPRP protein involved in polyphenol anchoring', *Plant Mol. Biol.*, 55, 763–80.

Blando, F., Spirito, R., Gerardi, C., Durante, M. and Nicoletti, I. (2012), 'Nutraceutical properties in organic strawberries from South Italy', *Acta Hort.*, 926, 683–90.

Borowy, A., Chrzanowska, E. and Kapłan, M. (2018), 'Comparison of three sour cherry cultivars grown in central-eastern Poland', *Acta Sci. Pol. Hortoru.*, 17(1), 63–73.

Boss, P. K., Davies, C. and Robinson, S. P. (1996), 'Expression of anthocyanin biosynthesis pathway genes in red and white grapes', *Plant Mol. Biol.*, 32(3), 565–9.

Brandt, K. and Mølgaard, J. P. (2001), 'Organic agriculture: Does it enhance or reduce the nutritional value of plant foods?' *J. Sci. Food Agric.*, 81, 924–31.

Britton, G. (2008), 'Functions of intact carotenoids', in Britton, G., Liaaen-Jensen, S. and Pfander, H. (Eds), *Carotenoids*, Basel: Birkhäuser, pp. 189–212.

Brovelli, E. A. (2006), 'Pre- and postharvest factors affecting nutraceutical properties of horticultural products', *Stewart Postharvest Rev.*, 2(2), 1–6.

Bruno, A., Durante, M., Marrese, P. P., Migoni, D., Laus, M., Pace, E., Pastore, D., Mita, G., Piro, G. and Lenucci, M. S. (2017), 'Shades of red: Comparative study on supercritical CO_2 extraction of lycopene-rich oleoresins from gac, tomato and watermelon fruits and effect of the α-cyclodextrin clathrated extracts on cultured lung adenocarcinoma cells' viability', *J. Food Comp. Anal.*, 65, 25–32.

Bureau, S., Renard, C. M. G. C., Reich, M., Ginies, C. and Audergon, J.-M. (2009), 'Change in anthocyanin concentrations in red apricot fruits during ripening', *LWT*, 42(1), 372–7.

Calvo, P., Nelson, L. and Kloepper, J. W. (2014), 'Agricultural uses of plant biostimulants', *Plant Soil*, 383, 3–41. doi:10.1007/s11104-014-2131-8.

Cantos, E., Espín, J. C. and Tomás-Barberán, F. A. (2002), 'Varietal differences among the polyphenol profiles of seven table grape cultivars studied by LC–DAD–MS–MS', *J. Agric. Food Chem.*, 50, 5691–6.

Castellanos-Morales, V., Villegas, J., Wendelin, S., Vierheilig, H., Eder, R. and Cárdenas-Navarro, R. (2010), 'Root colonisation by the arbuscular mycorrhizal fungus *Glomus intraradices* alters the quality of strawberry fruits (*Fragaria* × *ananassa* Duch.) at different nitrogen levels', *J. Sci. Food Agric.*, 90, 1774–82.

Castle, W. S. (1995), 'Rootstock as a fruit quality factor in citrus and deciduous tree crops', *New Zeal. J. Crop Hort. Sci.*, 23(4), 383–94. doi:10.1080/01140671.1995.9513914.

Cazzonelli, C. I. and Pogson, B. J. (2010), 'Source to sink: Regulation of carotenoid biosynthesis in plants', *Trends Plant Sci.*, 15, 266–74.

Cevallos-Casals, B. A., Byrne, D., Okie, W. R. and Cisneros-Zevallos, L. (2006). 'Selecting new peach and plum genotypes rich in phenolic compounds and enhanced functional properties', *Food Chem.*, 96, 273–328. doi:10.1016/j.foodchem.2005.02.032.

Cheng, J. H., Wei, L. Z. and Wu, J. (2015), 'Effect of light quality selective plastic films on anthocyanin biosynthesis in *Vitis vinifera* L. cv. Yatomi Rosa', *J. Agr. Sci. Tech.*, 17, 157–66.

Cordenunsi, B. R., Do Nascimento, J. R. O., Genovese, M. I. and Lajolo, F. M. (2002), 'Influence of cultivar on quality parameters and chemical composition of strawberry fruits grown in Brazil', *J. Agric. Food Chem.*, 50, 2581–6.

Corral-Aguayo, R. D., Yahia, E. M., Carrillo-Lopez, A. and González-Aguilar, G. (2008), 'Correlation between some nutritional components and the total antioxidant capacity measured with six different assays in eight horticultural crops', *J. Agric. Food Chem.*, 56(22), 10498–504.

Crisosto, C. H., Johnson, R. S., Luza, J. G. and Crisosto, G. M. (1994), 'Irrigation regimes affect fruit soluble solids concentration and rate of water loss of "O'Henry" peaches', *HortSci.*, 29, 1169–71.

Cronje, P. J. R., Barry, G. H. and Huysamer, M. (2011), 'Postharvest rind breakdown of "Nules Clementine" mandarin is influenced by ethylene application, storage temperature and storage duration', *Postharvest Biol. Technol.*, 60, 192–201.

Cronje, P. J. R., Barry, G. H. and Huysamer, M. (2013), 'Canopy position affects pigment expression and accumulation of flavedo carbohydrates of "Nules Clementine" mandarin fruit, thereby affecting rind condition', *J. Am. Soc. Hortic. Sci.*, 138, 217–24.

Davey, M. W. and Keulemans, J. (2004), 'Determining the potential to breed for enhanced antioxidant status in *Malus*: Mean inter- and intravarietal fruit vitamin C and glutathione contents at harvest and their evolution during storage', *J. Agric. Food Chem.*, 52(26), 8031–8.

Davies, M. B., Austin, J. and Partridge, D. A. (1991), *Vitamin C: Its Chemistry and Biochemistry*, Royal Society of Chemistry, Cambridge, UK, p. 150.

De Pascual-Teresa, S. and Sanchez-Ballesta, M. T. (2008), 'Anthocyanins: From plant to health', *Phytochem. Rev.*, 7, 281–99.

De Souza, V. R., Pereira, P. A. P., Silva, T. L. T., Lima, L. C. O. L., Pio, R. and Queiroz, F. (2014), 'Determination of the bioactive compounds, antioxidant activity and chemical composition of Brazilian blackberry, red raspberry, strawberry, blueberry and sweet cherry fruits', *Food Chem.*, 156, 362–8.

Delgado-Pelayo, R., Gallardo-Guerrero, L. and Hornero-Méndez, D. (2014), 'Chlorophyll and carotenoid pigments in the peel and flesh of commercial apple fruits varieties', *Food Res. Int.*, 65, 272–81.

Delva, L. (2012), 'Acerola (*Malphighia emarginata* DC): Phenolic profiling, antioxidant capacity, antimicrobial property, toxicological screening, and color stability', *PhD Thesis*, University of Florida, 174p.

Demir, T. (2013), 'Determination of carotenoid, organic acid and sugar content in some sweet cherry cultivars grown in Sakarya, Turkey', *J. Food Agric. Environ.*, 11, 73–5.

Di Matteo, A. D., Russo, R., Graziani, G., Ritieni, A. and Di Vaio, C. (2016), 'Characterization of autochthonous sweet cherry cultivars (*Prunus avium* L.) of southern Italy for fruit quality, bioactive compounds and antioxidant activity', *J. Sci. Food Agric.*, 97, 2782–94.

Dias, M. G., Camões, M. F. G. F. C. and Oliveira, L. (2009), 'Carotenoids in traditional Portuguese fruits and vegetables', *Food Chem.*, 113, 808–15.

Dixon, R. and Paiva, N. (1995), 'Stress-induced phenylpropanoid metabolism', *Plant Cell*, 7(7), 1085–97. doi:10.1105/tpc.7.7.1085.

Dorais, M. and Ehret, D. L. (2008), 'Agronomy and the nutritional quality of fruit', in Tomás-Barbéran, F. and Gil, I. M. (Eds), *Improving the Health- Promoting Properties of Fruit and Vegetable Products*, Cambridge, UK: Woodhead Publishing Limited, pp. 346–91.

Doumett, S., Fibbi, D., Cincinelli, A., Giordani, E., Nin, S. and Del Bubba, M. (2011), 'Comparison of nutritional and nutraceutical properties in cultivated fruits of *Fragaria vesca* L. produced in Italy', *Food Res. Intern.*, 44, 1209–16.

Du Jardin, P. (2015), 'Plant biostimulants: Definition, concept, main categories and regulation', *Sci. Hort.*, 196, 3–14. doi:10.1016/j.scienta.2015.09.021.

Dussi, M. C., Sugar, D. and Worlstad, R. E. (1995), 'Characterizing and quantifying anthocyanins in red pears and the effect of light quality on fruit color', *J. Am. Soc. Hort. Sci.*, 120, 785–9.

Erkan, M. and Kader, A. A. (2011), 'Pomegranate (*Punica granatum* L.)', in Yahia, E. M. (Ed.), *Postharvest Biology and Technology of Tropical and Subtropical Fruits: Mangosteen to White Sapote*, Cambridge, UK: Woodhead Publishing, pp. 287–311.

Esplay, R. V., Hellens, R. P., Putterill, J., Stevenson, D. E., Kutty-Amma, S. and Allan, A. C. (2007), 'Red colouration in apple fruit is due to the activity of the *MYB* transcription factor, *MdMYB10*', *Plant J.*, 49, 414–27. doi:10.1111/j.1365-313X.2006.02964.x.

Fanciullino, A. L., Cer os, M., Dhique-Mayer, C., Froelicher, Y., Talón, M., Ollitrault, P. and Morillon R. (2008), 'Changes in carotenoid content and biosynthetic gene expression in juice sacs of four orange varieties (*Citrus sinensis*) differing in flesh fruit color', *J. Agric. Food Chem.*, 56, 3628–38.

Feng, S., Wang, Y., Yang, S., Xu, Y. and Chen, X. (2010), 'Anthocyanin biosynthesis in pears is regulated by a R2R3-MYB transcription factor *PyMYB10*', *Planta*, 232, 245–55.

Gao, L. and Mazza, G. (1995), 'Characterization, quantitation, and distribution of anthocyanins and colorless phenolics in sweet cherries', *J. Agric. Food Chem.*, 43, 343–6.

Gest, N., Garchery, C., Gautier, H. Jiménez, A. and Stevens, R. (2013), 'Light-dependent regulation of ascorbate in tomato by a monodehydroascorbate reductase localized in peroxisomes and the cytosol', *Plant Biotechnol. J.*, 11, 344–54.

Gill, M. I., Tomás-Barberán, F. A., Hess-Pierce, B. and Kader A. A. (2002), 'Antioxidant capacities, phenolic compounds, carotenoids, and vitamin C contents of nectarine, peach, and plum cultivars from California', *J. Agric. Food Chem.*, 50, 4976–82.

Giménez, M. J., Serrano, M., Valverde, J. M., Martínez-Romero, D., Castillo, S., Valero, D. and Guillén, F. (2017), 'Preharvest salicylic acid and acetylsalicylic acid treatments preserve quality and enhance antioxidant systems during postharvest storage of sweet cherry cultivars', *J. Sci. Food Agric.*, 97, 1220–8.

Goggin, F. L., Avila, C. A. and Lorence, A. (2010), 'Vitamin C content in plants is modified by insects and influences susceptibility to herbivory', *Bioessays*, 32(9), 777–90.

Gonçalves, B., Moutinho-Pereira, J., Santos, A., Silva, A. P., Bacelar, E., Correia, C. and Rosa, E. (2006), 'Scion-rootstock interaction affects the physiology and fruit quality of sweet cherry', *Tree Physiol.*, 26(1), 93–104.

Gross, J. (Ed.) (1991), *Pigments in Vegetables: Chlorophylls and Carotenoids*, New York, NY: Van Nostrand, pp. 3–74

Hakkinen, S., Heinonen, M., Karenlampi, S., Mykkanen, H., Ruuskanen, J. and Torronen, R. (1999), 'Screening of selected flavonoids and phenolic acids in 19 berries', *Food Res. Int.*, 32, 345–53.

Halliwell, B. (1994), 'Free radicals, antioxidants, and human disease: Curiosity, cause, or consequence?' *The Lancet*, 344, 721–4.

Harborne, J. B. (1980), 'Plant phenolics', in Bell, E. A. and Charlwood, B. V. (Eds), *Encyclopedia of Plant Physiology, volume 8, Secondary Plant Products*, Berlin, Germany: Springer, pp. 329–402.

Haskell, M. J. (2013), 'Provitamin A carotenoids as a dietary source of vitamin A', in Tanumihardjo, S. A. (Ed.), *Carotenoids and Human Health. Nutrition and Health*, Totowa, NJ: Humana Press, pp. 249–60.

Huang, H. W., Wang, Y., Zhang, Z. H., Jiang, Z. W. and Wang, S. M. (2004), '*Actinidia* germplasm resources and kiwifruit industry in China', *HortSci.*, 39(6), 1165–72.

Iglesias, I., Salvia, J., Torguet, L. and Montserrat, R. (2005), 'The evaporative cooling effects of overtree microsprinkler irrigation on "Mondial Gala" apples', *Sci. Hort.*, 103, 267–87.

Intrigliolo, D. S. and Castel, J. R. (2010), 'Response of grapevine cv. "Tempranillo" to timing and amount of irrigation: Water relations, vine growth, yield and berry and wine composition', *Irrig. Sci.*, 28, 113–25. doi:10.1007/s00271-009-0164-1.

Iriti, M., Rossoni, M., Borgo, M. and Faoro, F. (2004), 'Benzothiadiazole enhances resveratrol and anthocyanin biosynthesis in grapevine, meanwhile improving resistance to *Botrytis cinerea*', *J. Agric. Food Chem.*, 52(14), 4406–13.

Iriti, M., Vitalini, S., Di Tommaso, G., D'Amico, S., Borgo, M. and Faoro, F. (2011), 'New chitosan formulation prevents grapevine powdery mildew infection and improves polyphenol content and free radical scavenging activity of grape and wine', *Aust. J. Grape Wine R.*, 17(2), 263–9. doi:10.1111/j.1755-0238.2011.00149.x.

Jaakola, L. (2013), 'New insight into the regulation of anthocyanin biosynthesis in fruits', *Trends Plant Sci.*, 18(9), 477–83.

Jaganath, I. B. and Zainal, A. (2016), 'Controlled environment for enhanced and consistent production of (poly)phenols in specialty crops', in Asaduzzaman, Md. (Ed.), *Controlled Environment Agriculture – Production of Specialty Crops Providing Human Health Benefits through Hydroponics*, New York, NY: Nova Science Publishers, pp. 1–32.

Jakopic, J., Slatnar, A., Štampar, F., Veberi R. and Simoncic, A. (2012), 'Analysis of selected primary metabolites and phenolic profile of "Golden Delicious" apples from four production systems', *Fruits*, 67 (5), 377–86.

Jia, D., Fan, L., Liu, G., Shen, J., Liu, C. and Yuan, T. (2011), 'Effects of genotypes and bagging practice on content of β-carotene in apple fruits', *J. Agric. Sci.*, 3, 196–202.

Jifon, J. L. and Lester, G. E. (2009), 'Foliar potassium fertilization improves fruit quality of field-grown muskmelon on calcareous soils in south Texas', *J. Sci. Food Agric.*, 89, 2452–60.

Josuttis, M., Carlen, C., Crespo, P., Nestby, R., Toldam-Andersen, T. B., Dietrich, H. and Krüger, E. (2012), 'A comparison of bioactive compounds of strawberry fruit from Europe affected by genotype and latitude', *J. Berry Res.*, 2, 73–95. doi:10.3233/JBR-2012-029.

Kallio, H., Yang, B. and Peippo, P. (2002), 'Effects of different origins and harvesting time on vitamin C, tocopherols, and tocotrienols in sea buckthorn (*Hippophaë rhamnoides*) berries', *J. Agric. Food Chem.*, 50, 6136–42.

Kanellis, A. and Manganaris, G. A. (2014), 'Antioxidants and bioactive compounds in fruits', in Nath, P., Bouzayen, M., Matoo, A. K. and Pech J. C. (Eds), *Fruit Ripening: Physiology, Signalling and Genomics*, Wallingford, UK: CAB International, pp. 99–126.

Kato, M. (2012), 'Mechanism of carotenoid accumulation in citrus fruit', *J. Jpn. Soc. Hort. Sci.*, 81, 219–33.

Kato, M., Ikoma, Y., Matsumoto, H., Sugiura, M., Hyodo, H. and Yano, M. (2004), 'Accumulation of carotenoids and expression of carotenoid biosynthetic genes during maturation in citrus fruit', *Plant Physiol.*, 134, 824–37.

Kevers, C., Falkowski, M., Tabart, J., Defraigne, J. O., Dommes, J. and Pincemail, J. (2007), 'Evolution of antioxidant capacity during storage of selected fruits and vegetables', *J. Agric. Food Chem.*, 55(21), 8596–603.

Khan, A. S. and Singh, Z. (2007), 'Methyl jasmonate promotes fruit ripening and improves fruit quality in Japanese plum', *J. Hort. Sci. Biotechnol.*, 82, 695–706.

Karppinen, K., Zoratti, L., Sarala, M., Carvalho, E., Hirsimäki, J, Mentula, H., Martens, S., Häggman, H. and Jaakola, L. (2016), 'Carotenoid metabolism during bilberry (*Vaccinium myrtillus* L.) fruit development under different light conditions is regulated by biosynthesis and degradation', *BMC Plant Biol.*, 16, 95. doi:10.1186/s12870-016-0785-5.

Kim, S.-H. (2003), 'Molecular cloning and analysis of anthocyanin biosynthesis genes preferentially expressed in apple skin', *Plant Sci.*, 165, 403–13.

Kim, I. J., Ko, K. C., Kim, C. S. and Chung, W. I. (2001), 'Isolation and characterization of cDNAs encoding β-carotene hydroxylase in *Citrus*', *Plant Sci.*, 161, 1005–10.

Kondo, S. and Inoue, K. (1997), 'Abscisic acid (ABA) and 1-aminocyclopropane-l-carboxylic acid (ACC) content during growth of "Satohnishiki" cherry fruit, and the effect of ABA and ethephon application on fruit quality', *J. Hortic. Sci.*, 72(2), 221–7. doi:10.1080/14620316.1997.11515509.

Kong, J.-M., Chia, L.-S., Goh, N.-K., Chia, T.-F. and Brouillard, R. (2003), 'Analysis and biological activities of anthocyanins', *Phytochem.*, 64(5), 923–33. doi:10.1016/S0031-9422(03)00438-2.

Koureh, O. K., Bakhshi, D., Pourghayoumi, M. and Majidian, M. (2018), 'Comparison of yield, fruit quality, antioxidant activity, and some phenolic compounds of white seedless grape obtained from organic, conventional, and integrated fertilization', *Int. J. Fruit Sci.*, 18, 1–12.

Kschonsek, J., Wolfram, T., Stöckl, A. and Böhm, V. (2018), 'Polyphenolic compounds analysis of old and new apple cultivars and contribution of polyphenolic profile to the *in vitro* antioxidant capacity', *Antioxidants*, 7, 20. doi:10.3390/antiox7010020.

Łata, B. and Wi ska-Krysiak, M. (2010), 'Cultivar and seasonal variation in bioactive compounds of highbush blueberry fruits (*Vaccinium corymbosum* L.)', *Folia Horticul.*, 22(1), 31–5.

Lado, J., Zacarías, L., Gurrea, A., Page, A., Stead, A. and Rodrigo, M. J. (2015), 'Exploring the diversity in *Citrus* fruit coloration to decipher the relationship between plastid ultrastructure and carotenoid composition', *Planta*, 242, 645–61.

Laribi, A. I., Palou, L., Intrigliolo, D. S., Nortes, P. A., Rojas-Argudo, C., Taberner, V., Bartual, J. and Pérez-Gago, M. B. (2013), 'Effect of sustained and regulated deficit irrigation on fruit quality of pomegranate cv. "Mollar de Elche" at harvest and during cold storage', *Agric. Water Manag.*, 125, 61–70.

Lee, S. K. and Kader, A. A. (2000), 'Preharvest and postharvest factors influencing vitamin C content of horticultural crops', *Postharvest Biol. Technol.*, 20(3), 207–20.

Lee, J., Dossett, M. and Finn, C. E. (2012), '*Rubus* fruit phenolic research: The good, the bad, and the confusing', *Food Chem.*, 130(4), 785–96.

Leong, S. Y. and Oey, I. (2012), 'Effects of processing on anthocyanins, carotenoids and vitamin C in summer fruits and vegetables', *Food Chem.*, 53, 1577–8.

Lo Piero, A. R. (2015), 'The state of the art in biosynthesis of anthocyanin and its regulation in pigmented sweet oranges (*Citrus sinensis* L. Osbeck), *J. Agric. Food Chem.*, 63, 4031–41.

Locato, V., Cimini, S. and De Gara, L. (2013), 'Strategies to increase vitamin C in plants: From plant defense perspective to food biofortification', *Front. Plant Sci*, 4, 152.

Manganaris, G. A., Goulas, V., Vicente, A. R. and Terry, L. A. (2014), 'Berry antioxidants: Small fruits providing large benefits', *J. Sci. Food Agric.*, 94(5), 825–33.

Manganaris, G. A., Goulas, V., Mellidou, I. and Drogoudi, P. (2018), 'Antioxidant phytochemicals in fresh produce: Exploitation of genotype variation and advancements in analytical protocols', *Front. Chem.*, 5, 95. doi:10.3389/fchem.2017.00095.

Martínez-Gil, A. M., Garde-Cerdán, T., Martínez, L., Alonso, G. L. and Salinas, M. R. (2011), 'Effect of oak extract application to Verdejo grapevines on grape and wine aroma', *J. Agric. Food Chem.*, 59, 3253–63.

Marty, I., Bureau, S., Sarkissian, G., Gouble, B., Audergon, J. M. and Albagnac, G. (2005), 'Ethylene regulation of carotenoid accumulation and carotenogenic gene expression in colour-contrasted apricot varieties (*Prunus armeniaca*)', *J. Exp. Bot.*, 56, 1877–86.

Mas, T., Susperregui, J., Berké, B, Chèze, C., Moreau, S., Nurich, A. and Vercauteren, J. (2000), 'DNA triplex stabilization property of natural anthocyanins', *Phytochem.*, 53, 679–87.

Mazza, G. and Miniati, E. (1993), *Anthocyanins in Fruits, Vegetables and Grains*, Boca Raton, FL, USA: CRC Press.

Mazza, G. and Velioglu, Y. S. (1992), 'Anthocyanins and other phenolic compounds in fruits of red-flesh apples', *Food Chem.*, 43(2), 113–17. doi:10.1016/0308-8146(92)90223-O.

Mellidou, I. and Kanellis, A. K. (2017), 'Genetic control of ascorbic acid biosynthesis and recycling in horticultural crops', *Front. Chem.*, 5, 50. doi:10.3389/fchem.2017.00050.

Meyers, K. J., Mares, J. A., Igo, R. P., Truitt, B., Liu, Z., Millen, A. E., Klein, M., Johnson, E. J., Engelman, C. D., Karki, C. K., Blodi, B., Gehrs, K., Tinker, L., Wallace, R., Robinson, J., LeBlanc, E. S., Sarto, G., Bernstein, P. S., SanGiovanni, J. P. and Iyengar, S. K. (2014), 'Genetic evidence for role of carotenoids in age-related macular degeneration in the carotenoids in age-related eye disease study (CAREDS)', *Invest. Ophthalmol. Vis. Sci.*, 55(1), 587–99.

Milani, A., Basirnejad, M., Shahbazi, S. and Bolhassani, A. (2017), 'Carotenoids: Biochemistry, pharmacology and treatment', *Brit. J. Pharmacol.*, 174(11), 1290–324.

Miller, V., Mente, A., Dehghan, M., Rangarajan, S., Zhang, X., Swaminathan, S., Dagenais, G., Gupta, R., Mohan, V., Lear, S., Bangdiwala, S. I., Schutte, A. E., Wentzel-Viljoen, E., Avezum, A., Altuntas, Y., Yusoff, K., Ismail, N., Peer, N., Chifamba, J., Diaz, R., Rahman, O., Mohammadifard, N., Lana, F., Zatonska, K., Wielgosz, A., Yusufali, A., Iqbal, R., Lopez-Jaramillo, P., Khatib, R., Rosengren, A., Kutty, V. R., Li, W., Liu, J., Liu, X., Yin, L., Teo, K., Anand, S., Yusuf, S., Prospective Urban Rural Epidemiology (PURE) study investigators (2017), 'Fruit, vegetable, and legume intake, and cardiovascular disease and deaths in 18 countries (PURE): A prospective cohort study', *The Lancet*, 390(10107), 2037–49. doi:10.1016/S0140-6736(17)32253-5.

Mori, K., Sugaya, S. and Gemma, H. (2005), 'Decreased anthocyanin biosynthesis in grape berries grown under elevated night temperature condition', *Sci. Hort.*, 105, 319–30. doi:10.1016/j.scienta.2005.01.032.

Naik, P. S., Chanemougasoundharam, A., Paul Khurana, S. M. and Kalloo, G. (2003), 'Genetic manipulation of carotenoid pathway in higher plants', *Curr. Sci.*, 85, 1423–30.

Namitha, K. K. and Negi, P. S. (2010), 'Chemistry and biotechnology of carotenoids', *Crit. Rev. Food Sci.*, 50, 728–60.

Nollet, L. M. L. and Gutierrez-Uribe, J. A. (2018), *Phenolic Compounds in Food Characterization and Analysis*, Boca Raton, FL: CRC Press, p. 460.

Olas, B. (2018), 'Berry phenolic antioxidants – implications for human health?' *Front. Pharmacol.*, 9, 78.

Olsson, M. E., Andersson, C. S., Oredsson, S., Berglund, R. H. and Gustavsson, K-E. (2006), 'Antioxidant levels and inhibition of cancer cell proliferation in vitro by extracts from organically and conventionally cultivated strawberries', *J. Agric. Food Chem.*, 54(4), 1248–55.

Oyebode, O., Gordon-Dseagu, V., Walker, A. and Mindell, J. S. (2013), 'Fruit and vegetable consumption and all-cause, cancer and CVD mortality: Analysis of Health Survey for England data', *J. Epidemiol. Commun. Health*, 68, 856–62. doi:10.1136/jech-2013-203500.

Padayatty, S. J. and Levine, M. (2016), 'Vitamin C physiology: The known and the unknown and Goldilocks', *Oral Dis.* 22(6), 463–93.

Paliyath, G., Schofield, A., Oke, M. and Taehyun, A. (2002), 'Phosphorus fertilization and biosynthesis of functional food ingredients', in T. W. Bruulsema (Ed.), *Fertilizing Crops for Functional Foods*. Symposium Proceedings, 11 November 2002, Indianapolis, Indiana, USA. Potash and Phosphate Institute/Potash and Phosphate Institute of Canada, Section 5, pp. 5–6.

Pardo-García, A. I., Martínez-Gil, A. M., Cadahía, E., Pardo, F., Alonso, G. L. and Salinas, M. R. (2014), 'Oak extract application to grapevine as a plant biostimulant to increase wine polyphenols', *Food Res. Int.*, 55, 150–60. doi:10.1016/j.foodres.2013.11.004.

Parr, A. J. and Bowell, G. P. (2000), 'Phenols in the plant and in man: The potential for possible nutritional enhancement of the diet by modifying the phenols content or profile', *J. Sci. Food Agric.*, 80, 985–1012.

Patil, B. S., Crosby, K., Byrne, D. and Hirschi, K. (2014), 'The intersection of plant breeding, human health, and nutritional security: Lessons learned and future perspectives', *HortSci.*, 49(2), 116–27.

Pérez-Pastor A., Ruiz-Sànchez, M. C., Martinez, J. A., Nortes, P. A., Artes, F. and Domingo, R. (2007), 'Effect of deficit irrigation on apricot fruit quality at harvest and during storage', *J. Sci. Food Agric.*, 87, 2409–15.

Poiroux-Gonord, F., Bidel, L. P. R., Fanciullino, A.-L., Gautier, H., Lauri-Lopez, F. and Urban, L. (2010), 'Health benefit of vitamins and secondary metabolites of fruits and vegetables and prospects to increase their concentrations by agronomic approaches', *J. Agric. Food Chem.*, 58, 12065–82. doi:10.1021/jf1037745.

Rahman, M. R., Sabir, A. A., Mukta J. A., Khan M. M. A., Mohi-Ud-Din M., Miah, M. G., Rahman M. and Islam T. (2018), 'Plant probiotic bacteria *Bacillus* and *Paraburkholderia* improve growth, yield and content of antioxidants in strawberry fruit', *Sci. Rep.*, 8, 2504.

Ramful, D., Tarnus, E., Aruoma, O. I., Bourdon, E. and Bahorun, T. (2011), 'Polyphenol composition, vitamin C content and antioxidant capacity of Mauritian citrus fruit pulps', *Food Res. Intern.*, 44(7), 2088–99.

Rapisarda, P., Bianco, M. L., Pannuzzo, P. and Timpanaro, N. (2008), 'Effect of cold storage on vitamin C, phenolics and antioxidant activity of five orange genotypes [*Citrus sinensis* (L.) Osbeck]', *Postharvest Biol. Technol.*, 49(3), 348–54.

Reboul, E., Richelle, M., Perrot, E., Desmoulins-Malezet, C., Pirisi, V. and Borel, P. (2006), 'Bioaccessibility of carotenoids and vitamin E from their main dietary sources', *J. Agric. Food Chem.*, 54, 8749–55.

Reganold, J. P., Andrews, P. K., Reeve, J. R., Carpenter-Boggs, L., Schadt, C. W., Alldredge, J. R., Ross, C. F., Davies, N. M. and Zhou, J. (2010), 'Fruit and soil quality of organic and conventional strawberry agroecosystems', *PLoS ONE*, 5(9), e12346. doi:10.1371/journal.pone.0012346.

Remorini, D., Tavarini, S., Degl'Innocenti, E., Loreti, F., Massai, R. and Guidi, L. (2008), 'Effect of rootstocks and harvesting time on the nutritional quality of peel and flesh of peach fruits', *Food Chem.*, 110, 361–7. doi:10.1016/j.foodchem.2008.02.011.

Rodrigo, M. J., Marcos, J. F. and Zacarías, L. (2004), Biochemical and molecular analysis of carotenoid biosynthesis in flavedo of orange (*Citrus sinensis* L.) during fruit development and maturation', *J. Agric. Food Chem.*, 52, 6724–31.

Rodrigo, M. J., Alquézar, B., Alós E., Medina V., Carmona L., Bruno M., Al-Babili S. and Zacarías, L. (2013), 'A novel carotenoid cleavage activity involved in the biosynthesis of *Citrus* fruit-specific apocarotenoid pigments', *J. Exp. Bot.*, 64(14), 4461–78.

Rodriguez-Amaya, D. B. (2001), *A Guide to Carotenoid Analysis in Foods*, Washington DC, USA: ILSI Press, pp. 41–3.

Ruiz, D., Egea, J., Tomás-Barberán, F. A. and Gil, M. I. (2005a), 'Carotenoids from new apricot (*Prunus armeniaca* L.) varieties and their relationship with flesh and skin colour', *J. Agric. Food Chem.*, 53, 6368–74.

Ruiz, D., Egea, J., Gil, M. I. and Tomás-Barberán, F. A. (2005b), 'Characterization and quantitation of phenolic compounds in new apricot (*Prunus armeniaca* L.) varieties', *J. Agric. Food Chem.*, 53(24), 9544–52. doi:10.1021/jf051539p.

Ruiz-García, Y. and Gómez-Plaza, E. (2013), 'Elicitors: A tool for improving fruit phenolic content', *Agriculture*, 3(1), 33–52.

Russo, M., Moccia, S., Bilotto, S., Spagnuolo, C., Durante, M., Lenucci, M. S., Mita, G., Volpe, M. G., Aquino, R. P. and Russo, G. L. (2017), 'A carotenoid extract from a Southern Italian cultivar of pumpkin triggers nonprotective autophagy in malignant cell', *Oxid. Med. Cell. Longev.* doi:10.1155/2017/7468538.

Sánchez-Gómez, R., Zalacain, A., Pardo, F., Alonso G. L. and Salinas, M. R. (2017), 'Moscatel vine-shoot extracts as grapevine biostimulant to enhance wine quality', *Food Res. Int.*, 98, 40–9. doi:10.1016/j.foodres.2017.01.004.

Sarma, A. D. and Sharma, R. (1999), 'Anthocyanin-DNA copigmentation complex: Mutual protection against oxidative damage', *Phytochem.*, 52(7), 1313–18.

Sass-Kiss, A., Kiss, J., Milotay, P., Kerek, M. M. and Toth-Markus, M. (2005), 'Differences in anthocyanin and carotenoid content of fruits and vegetables', *Food Res. Int.*, 38, 1023–9.

Sayyari, M., Valero, D., Babalar, M., Kalantari, S., Zapata, P. J. and Serrano, M. (2010), 'Pre-storage oxalic acid treatment maintained visual quality, bioactive compounds, and antioxidant potential of pomegranate after long-term storage at 2°C', *J. Agric. Food Chem.*, 58(11), 6804–8.

Scalzo, J., Politi, A., Pellegrini, N., Mezzetti, B. and Battino, M. (2005), 'Plant genotype affects total antioxidant capacity and phenolic contents in fruit', *Nutrition*, 21, 207–13. doi:10.1016/j. nut.2004.03.025.

Sharoni, Y., Linnewiel-Hermoni, K., Khanin, M., Salman, H., Veprik, A., Danilenko, M. and Levy, J. (2012), 'Carotenoids and apocarotenoids in cellular signaling related to cancer: A review', *Mol. Nutr. Food Res.*, 56, 259–69.

Singh, K, P., Singh, A. and Singh, U. P. (2015), 'Phenolic acid content of some apple cultivars with varying degrees of resistance to apple scab', *Int. J. Fruit Sci.*, 15(3), 267–80.

Sivakumaran, S., Huffman, L., Sivakumaran, S. and Drummond, L. (2018), 'The nutritional composition of Zespri® SunGold kiwifruit and Zespri® Sweet Green kiwifruit', *Food Chem.*, 238, 195–202.

Skibsted, L. H. (2012), 'Carotenoids in antioxidant networks colorants or radical scavengers', *J. Agric. Food Chem.*, 60, 2409–17.

Slatnar, A., Petkovsek, M. M., Halbwirth, H., Štampar, F., Stich, K. and Veberi , R. (2010), 'Response of the phenylpropanoid pathway to *Venturia inaequalis* infection in maturing fruit of "Braeburn" apple', *J. Hort. Sci. Biotechnol.*, 85, 465–72.

Stinco, C. M., Baroni, M. V., Naranjo, R. D. D. P., Wunderlin, D. A., Heredia, F. J., Meléndez-Martínez, A. J. and Vicario, I. M. (2015), 'Hydrophilic antioxidant compounds in orange juice from different fruit cultivars: Composition and antioxidant activity evaluated by chemical and cellular based (*Saccharomyces cerevisiae*) assays', *J. Food Comp. Anal.*, 37, 1–10.

Stephensen, C. B. (2013), 'Provitamin: A carotenoids and immune function', *in* S. A. Tanumihardjo (Ed.), *Carotenoids and Human Health, Nutrition and Health*. Humana Press, Totowa, NJ, pp. 261–70.

Steyn, W. J., Wand, S. J. E., Holcroft, D. M. and Jacobs, G. (2002), 'Anthocyanins in vegetative tissues: A proposed unified function in photoprotective', *New Phytol.*, 155, 349–61.

Steyn, W. J., Wand, S. J. E., Holcroft, D. M. and Jacobs, G. (2005), 'Red colour development and loss in pears', *Acta Hort.*, 671, 79–85.

Steyn, W., Wand, S. J. E., Jacobs, G., Rosecrance, R. C. and Roberts, S. C. (2009), 'Evidence for a photoprotective function of low-temperature-induced anthocyanin accumulation in apple and pear peel', *Physiol. Plant.*, 136, 461–72. doi:10.1111/j.1399-3054.2009.01246.x.

Sun, J., Chu, Y. F., Wu, X. and Liu, R. H. (2002), 'Antioxidant and antiproliferative activities of common fruits', *J. Agric. Food Chem.*, 50(25), 7449–54.

Suza, W. P., Avila, C. A., Carruthers, K., Kulkarni, S., Goggin, F. L. and Lorence, A. (2010), 'Exploring the impact of wounding and jasmonate on ascorbate metabolism', *Plant Physiol. Biochem.*, 48(5), 337–50.

Tavarini, S., Gil, M. I., Tomás-Barberán, F. A., Buendia, B., Remorini, D., Massai, R., Degl'Innocenti, E. and Guidi, L. (2011), 'Effects of water stress and rootstocks on fruit phenolic composition and physical/chemical quality in Suncrest peach', *Ann. Appl. Biol.*, 158(2), 226–33.

Teleszko, M., Wojdyło, A., Rudzin ska, M., Oszmian ski, J. and Golis, T. (2015), 'Analysis of lipophilic and hydrophilic bioactive compounds content in sea buckthorn (*Hippophae rhamnoides* L.) berries', *J. Agric. Food Chem.*, 63(16), 4120–9.

Tomás-Barberán F. A., Gil, M. I., Cremin, P., Waterhouse, A. L., Hess-Pierce, B. and Kader, A. A. (2001), 'HPLC-DAD-ESIMS analysis of phenolic compounds in nectarines, peaches and plums', *J. Agric. Food Chem.*, 49, 4748–60.

Torrecillas A, Domingo R., Galego R. and Ruiz-Sánchez M. C. (2000), 'Apricot tree response to withholding irrigation at different phenological periods', *Sci. Hortic.*, 85, 201–15.

Treutter, D. (2010), 'Managing phenol contents in crop plants by phytochemical farming and breeding – visions and constraints', *Int. J. Mol. Sci.*, 11, 807–57.

Tsuda, T., Shiga, K., Ohshima, K., Kawakishi, S. and Osawa, T. (1996), Inhibition of lipid peroxidation and the active oxygen radical scavenging effect of anthocyanin pigments isolated from *Phaseolus vulgaris* L., *Biochem. Pharmacol.*, 52, 1033–9.

Usenik, V., Štampar, F. and Veberi , R. (2009), 'Anthocyanins and fruit colour in plums (*Prunus domestica* L.) during ripening', *Food Chem.*, 114(2), 529–34. doi:10.1016/j.foodchem.2008.09.083.

Valero, D. (2015), 'Recent developments to maintain overall sweet cherry quality during postharvest storage', *Acta Hort.*, 1079, 83–94.

Veberi , R. (2016), 'The impact of production technology on plant phenolics', *Horticulturae*, 2, 8. doi:10.3390/horticulturae2030008.

Wang, D., Wang, X., Zhang, C., Ma, Y. and Zhao, X. (2011), 'Calf thymus DNA-binding ability study of anthocyanins from purple sweet potatoes (*Ipomea batatas* L.)', *J. Agric. Food Chem.* 50(13), 7405–9.

Yang, X.-Y., Wang, F.-F., Teixeira da Silva, J. A., Zhong, J., Liu, Y.-Z. and Peng, S.-A. (2013), Branch girdling at fruit green mature stage affects fruit ascorbic acid contents and expression of genes involved in L-galactose pathway in citrus. *New Zeal. J. Crop Hort. Sci.*, 41(1), 23–31.

Yoo, K. M. and Moon, B. K. (2016), 'Comparative carotenoid compositions during maturation and their antioxidative capacities of three citrus varieties', *Food Chem.*, 196, 544–9.

Young, J. I., Züchner, S. and Wang, G. (2015), 'Regulation of the epigenome by vitamin C', *Annu. Rev. Nutr.*, 35(1), 545–64.

Zainudin, M. A. M., Hamid, A. A., Anwar, F., Osman, A. and Saari, N. (2014), 'Variation of bioactive compounds and antioxidant activity of carambola (*Averrhoa carambola* L.) fruit at different ripening stages', *Sci. Hort.*, 172, 325–31.

Zhang, L., Ma, G., Kato, M., Yamawaki, K., Takagi, T., Kiriiwa, Y., Ikoma, Y., Matsumoto, H., Yoshioka, T. and Nesumi, H. (2012), 'Regulation of carotenoid accumulation and the expression of carotenoid metabolic genes in citrus juice sacs *in vitro*', *J. Exp. Bot.*, 63, 871–86.

Zhu, L., Zhang Y. and Lu, J. (2012), 'Phenolic contents and compositions in skins of red wine grape cultivars among various genetic backgrounds and originations', *Int. J. Mol. Sci.*, 13, 3492–510.

Zoratti, L., Karppinen, K., Luengo Escobar, A., Häggman, H. and Kaakola, L. (2014a), 'Light-controlled flavonoid biosynthesis in fruits', *Front. Plant Sci.*, 5, 534. doi:10.3389/fpls.2014.00534.

Zoratti, L., Sarala, M., Carvalho, E., Karppinen, K., Martens, S., Giongo, L., Häggman, H. and Kaakola, L. (2014b), 'Monochromatic light increases anthocyanin content during fruit development in bilberry', *BMC Plant Biol.*, 14, 377. doi:10.1186/s12870-014-0377-1.

Chapter 2

The nutritional and nutraceutical/functional properties of mangoes

Laurent Urban, University of Avignon, France; Mônica Maria de Almeida Lopes; and Maria Raquel Alcântara de Miranda, Federal University of Ceará, Brazil

1 Introduction

Mangoes are recognized as a major source of bioactive compounds with potential health-promoting activities (thereafter named phytochemicals). We review here the potential health benefits of mango fruits that can be derived from what we know about the biological effects of vitamin C, carotenoids and phenolic compounds and from what we learned from more specific studies performed on cell and animal models. An up-to-date list of the pre- and post-harvest factors influencing the concentrations in phytochemicals of the pulp of mango fruits is presented. There are not many observations about pre-harvest treatments. By contrast, the literature is especially abundant on post-harvest treatments, including, besides coating techniques, exposure to ionizing radiations; electrical fields; visible, ultraviolet (UV) or infrared (IR) radiations, either in a 'traditional' form or under the form of pulsed light (PL); heat or cold; high pressure and chemicals such as salicylic acid. It is quite obvious from this list that almost all these treatments may be considered as stressing or mimicking stress (salicylic acid is involved in biotic stress signalling). We shall therefore evoke some of the current hypotheses about the stimulating effect of stress on the secondary metabolism.

It is important to keep in mind that all parts of mango trees have been used mainly in traditional South Asian, South American and African medicine (Masibo and He, 2008). However, in this chapter, we shall rather put emphasis on the pulp of either entire fruits or fresh cuts, which represent the forms that are currently available to consumers. For reviews

http://dx.doi.org/10.19103/AS.2017.0026.17

about the potentials and uses of by-products of the food industry, see Jahurul et al. (2015), Masibo and He (2008) and Ribeiro and Schieber (2010).

2 Health benefits of mango fruits

Mango pulp and crude natural extracts from the peel or the pulp are rich in various phytochemicals, including phenolics such as gallic acid-O-hexoside, syringic acid hexoside, gallic acid (GA; Septembre-Malaterre et al., 2016) and glycosylated xanthine (mangiferin) (Oliveira et al., 2016); flavonols such as quercetin, naringenin, kaempferol, apigenin, luteolin and chrysin (Pierson et al., 2014); carotenoids such as β-carotene (Mercadante et al., 1997) and vitamin C (Tian et al., 2010) (Table 1). Most of these compounds have antioxidant properties and contribute to the antioxidant capacity of mango fruits, measured by oxygen radical absorbance capacity (ORAC), ferric reducing antioxidant power (FRAP), (2.2-diphenyl-1-picrylhydrazyl) (DPPH) and 2,2-azinobis (3-ethyl-benzothiazoline-6-sulphonic acid) (ABTS) tests.

Table 1 Range of contents for the major classes of phytochemicals and major compounds found in mango pulp

Phytochemicals	Range of contents	References
Vitamin C	9.8 to 186 mg 100 g^{-1} fresh weight	Nisperos-Carriedo et al. (1992) Vinci et al. (1995) Franke et al. (2004) Gil et al. (2006) Reyes and Cisneros-Zevallos (2007) Ribeiro et al. (2007) Corral-Aguayo et al. (2008) Manthey and Perkins-Veazie (2009) Tian et al. (2010)
Total phenolics	9 to 208 mg 100 g^{-1} fresh weight	Gil et al. (2006) Ribeiro et al. (2007) Manthey and Perkins-Veazie (2009)
Gallic acid	69 mg 100 g^{-1} fresh weight	Schieber et al. (2000)
Mangiferin	3.0 to 19.4 mg kg^{-1} dry basis	Berardini et al. (2005)
Total carotenoids	1159-ca. 3000 mg 100 g^{-1} fresh weight	Mercadante et al. (1997) Hulshof et al. (1997) Ben-Amotz and Fishier (1998) Setiawan et al. (2001) Pott et al. (2003) Chen et al. (2004) Ornelas-Paz et al. (2007) Ribeiro et al. (2007) Veda et al. (2007) Corral-Aguayo et al. (2008)
β-carotene	0.55–3.21 mg 100 g^{-1} fresh weight	
Antioxidant capacity	6.12 to 81.39	Shi et al. (2015)

2.1 Theoretical health benefits of mango phytochemicals

Vitamin C, also known as ascorbate, is a vital micronutrient for humans. A lack of vitamin C hampers the activity of a range of enzymes and may lead to scurvy in humans (Olmedo et al., 2006). Unlike most animals, humans are unable to synthesize their own vitamin C, and they must therefore find it in plants, in particular, fruits and vegetables (FAVs). In addition to its involvement in the production of collagen, ascorbic acid serves as a cofactor in several vital enzymatic reactions, including those involved in the synthesis of catecholamines, carnitine and cholesterol, and in the regulation of transcription factors controlling the expression of important genes of the metabolism (Arrigoni and De Tullio, 2002). Ascorbic acid is present in three forms: ascorbate, monodehydroascorbate and dehydroascorbate which corresponds to the oxidized form of ascorbate. In most cellular functions, ascorbate acts as an electron donor, but it may also act directly to scavenge reactive oxygen species (ROS) generated by cellular metabolism. Due to the role of ascorbate in protecting cells against oxidative stress and the involvement of ROS in neurodegenerative disorders (Alzheimer's and Parkinson's diseases) or inflammatory response (atherosclerosis), it is strongly suggested that vitamin C plays a positive role in the prevention of heart, chronic inflammatory and neurodegenerative diseases (Ames et al., 1993).

Gallic acid is a phenolic acid, 3, 4, 5-trihydroxybenzoic acid. Its dimeric derivative is known as ellagic acid. Both GA and ellagic acid exist either in the free or in the bound form. The bound forms of GA and of ellagic acid are called gallotannins (GTs) and ellagitannins, respectively. These two molecules can react with one another to form diagallic acid, which is an ester. GA is the major compound among the phenolic acids found in the mango pulp (Schieber et al., 2000). GA was shown to have antioxidant, anti-inflammatory, antimicrobial, anti-mutagenic, anticancer and free-radical-scavenging properties (Madsen and Bertelsen, 1995).

Mangiferin is a xanthone, C-2-β–ᴅ–glucopyranosyl-1, 3, 6, 7-tetrahydroxyxanthone. Xanthones are thought to be among the most powerful antioxidants known (Masibo and He, 2008). Mangiferin is a pharmacologically active phytochemical, with antioxidant, anti-inflammatory, antimicrobial, anti-atherosclerotic, anti-allergenic, analgesic and immunomodulary properties. See Masibo and He (2008) for a detailed review.

Quercetin is a flavonoid, often occurring in plants as glycosides such as rutin. The predominant flavonol glycoside found in mango pulp is quercetin 3-galactoside, followed by quercetin 3-arabinoside according to Schieber et al. (2000). It is, however, important to keep in mind that there are many sources of quercetin in FAVs and that quercetin supply by mango pulp exists. Other flavonol glycosides, like kaempferol, were found only in trace amounts in mango (Schieber et al., 2000). Quercetin has been found to have antihistamine, anti-inflammatory and anticancer properties, and helps prevent cardiovascular diseases. Also quercetin is thought to contribute to protect against chronic diseases such as diabetes, obesity, atherosclerosis and heart disease.

On absorption, quercetin is metabolized mainly to isorhamnetin, tamarixetin and kaempferol. Kaempferol is a strong antioxidant. It inhibits monocyte chemoattractant protein MCP-1 which plays a role in the initial steps of atherosclerotic plaque formation. See Chen and Chen (2013) for a recent review about the numerous health benefits of kaempferol.

Carotenoids endowed with provitamin A activity are vital components of the human diet. Vitamin A is involved in hormone synthesis, immune responses, and the regulation of cell growth and differentiation (Combs, 1995). It can be produced within certain tissues from carotenoids such as β-carotene which is present in the pulp of mango fruits. A carotenoid-deficient diet can lead to night blindness and premature death. Carotenoid-rich diets are correlated with a significant reduction in the risk for certain cancers, coronary heart disease and several degenerative diseases. Carotenoids have demonstrated anticancer and anti-mutagenic properties (Krinsky and Johnson, 2005). Underlying mechanisms are not well understood, but the dietary importance of carotenoids is discussed, at least in part, in terms of antioxidant properties (Beutner et al., 2001; Combs, 1995; Krinsky and Johnson, 2005). Carotenoids are known for their capacity to efficiently quench $^1O_2$ singlet oxygen by energy transfer (Baltschun et al., 1997). $^1O_2$ is a particularly active ROS, capable of damaging DNA (Sies and Menck, 1992) and provoking genetic mutations (Devasagayam et al., 1991). Eventually, $^1O_2$ can damage lipids and membranes (Kalyanaraman et al., 1987). β-carotene is an efficient antioxidant, capable of inhibiting strongly the formation of peroxide. β-carotene is prone to degradation after ingestion, but its breakdown product seems to have interesting properties that may explain the cancer preventative activity (Linnewiel et al., 2009). When lipophilic antioxidants such as lutein or lycopene are associated with hydrophilic antioxidants such as rutin, a supra-additive protection of low-density lipoprotein occurs (Mildel et al., 2007). When rutin is associated with ascorbic acid, a synergetic protection also occurs.

2.2 Mango antioxidants

As said before, many of the above-mentioned compounds are antioxidants. When the compounds endowed with antioxidant capacity are adequately released from the food matrix and absorbed in the small intestine (see below), they are believed to protect the different body tissues against oxidative stress, conferring different health benefits (Das et al., 2012; Palafox-Carlos et al., 2011). Table 2 summarizes the effects of pre- and post-harvest factors on the antioxidant properties of mango pulp.

The antioxidant capacity of phenolics is attributed mainly to the number and localization of their hydroxyl groups and their interactions with dietary fibres (DFs) which could not only limit their absorption, but also prevent these groups from stabilizing free radicals (Palafox-Carlos et al., 2011). Velderrain-Rodríguez et al. (2016) evaluated the effect of DF present in 'Ataulfo' mangoes on bioaccessibility (see below) of phenolics and antioxidant capacity in an 'in vitro' digestion model that simulates the conditions of the human gastrointestinal tract. They concluded that DF did not represent a major limitation to the bioaccessibility of phenolics (2.48 mg GAE 100 g^{-1} FW) in mangoes.

There is, however, a debate. It would certainly be wrong to attribute the health benefits of FAVs solely to the antioxidant properties of the phytochemicals they supply, even if many of the diseases they contribute to prevent involve oxidative and inflammatory stress. The term 'antioxidant paradox' is used to refer to the observation that giving large doses of dietary antioxidant supplements to human consumers demonstrates, in most studies, little or no preventative or therapeutic effect even for diseases in which ROS are important (Halliwell, 2013). The explanation for the positive effects of phytochemicals may be that, besides their antioxidant properties, many of them may act as elicitors that alter transcription, among others activate Nrf2, a transcription factor that binds to the antioxidant response element in the promoter region of genes coding for enzymes involved in protective mechanisms (Surh and Na, 2008). In addition, there is now mounting evidence that phytochemicals, such as phenolics, may act indirectly through the mediation of the gut microbiota (Anhê et al., 2015).

Table 2 Influence of pre- and post-harvest factors on antioxidant activities in mango pulp, measured by different methods. See the text for a definition of the ripening stages (RS)

Mango (Cultivar)	Factors	Major findings	Antioxidant activity assay	References
'Keitt'	Ripening stage	High correlation between antioxidant activity and total phenolics. Ripening stages RS2 and RS6 characterized by high concentrations of phytochemicals	ORAC[1]	Ibarra-Garza et al. (2015)
'Samar Bahisht Chaunsa'	Ripening stage	Increase in antioxidant activity up to day 7	DPPH[2]	Razzaq et al. (2013)
'Ataulfo'	Ripening stage	Ripening stages RS2 and RS3 characterized by an increased antioxidant capacity	FRAP[3] and DPPH	Palafox-Carlos et al. (2012)
28 mango genotypes	Genetic factors	Total antioxidant potency composite index varied among all genotypes from 6.12 to 81.39, and was significantly correlated with total phenolics	ABTS[4], FRAP, MCC[5], ABTS, SRSA[6] and DPPH	Shi et al. (2015)
'Tommy Atkins'	Cultural practice (biodynamic culture, organic and conventional systems)	Highest antioxidant activity observed in mature green and ripe mango fruits from biodynamic culture	DPPH	Maciel et al. (2010)
'Chok Anan'	Arabic gum plus calcium chloride	High DPPH radical scavenging activity maintained during low temperature storage	DPPH	Khaliq et al. (2016)
'Tainung'	UV-B exposure	Correlation between DPPH scavenging activity and total polyphenols and vitamin C	DPPH and FRAP	Jiang et al. (2015)
'Nam Dok Mai'	Cold storage and salicylic acid	Increased levels of vitamin C acid, total phenolics and antioxidant activity	DPPH and ABTS	Junmatong et al. (2015)

(Continued)

Table 2 (Continued)

Mango (Cultivar)	Factors	Major findings	Antioxidant activity assay	References
'Amrapali'	High pressure	Retention of 92% of total phenolics and 90% of antioxidant activity	DPPH and ABTS	Kaushik et al. (2014)
'Kent'	Fresh cut dipped in antioxidant solutions	Higher antioxidant activity observed as a consequence of the treatment using ascorbic acid at 1%	DPPH and ABTS	Robles-Sánchez et al. (2009)

[1] ORAC: oxygen radical absorbance capacity
[2] DPPH: (2.2-diphenyl-1-picrylhydrazyl)
[3] FRAP: ferric reducing antioxidant power
[4] ABTS: 2,2-azinobis (3-ethyl-benzothiazoline-6-sulphonic acid)
[5] MCC: metal chelating capacity
[6] SRSA: superoxide radical scavenging activity

2.3 Bioaccessibility and bioavailability

The term 'bioaccessibility' refers to the proper release of nutrients or specific phytochemicals within the food matrix as influenced by the conditions of the gastrointestinal tract (Saura-Calixto et al., 2007). Bioavailability refers to the total amount that is released and absorbed, reaching the bloodstream, where bioactive compounds are delivered to the different body tissues (Manach et al., 2005).

Although there is compelling evidence that vitamins and secondary metabolites are essential for human health, because they act either as antioxidants or through other mechanisms, many questions remain unresolved. Biologically active substances found in FAVs always come as part of a mixture in the diet. In a mixture, metabolites may have potentiating, antagonizing or synergistic effects (Raskin and Ripoll, 2004). Moreover, health benefits may be influenced by other ingredients such as DF, monounsaturated fatty acids, agents stimulating the immune system, minerals and even ethanol (Halliwell, 2007). Then, there is the issue of bioavailability of biologically active substances, which is affected by several factors such as tannin and lignin concentrations that differ greatly from one species to another.

Tannins have anti-feeding effects, due to their protein-binding properties, whereas lignin decreases the digestibility of plant material. Besides, it is now established that not all individuals respond identically to bioactive food components because of the existence of genetic profiles that modulate the responses. Finally, little is known about the dynamics of food components after they are ingested and then metabolized in the body.

On the basis of the above-mentioned questions and debates, it appears unfortunately not sufficient to demonstrate that mangoes supply large amounts of compounds theoretically known for being antioxidants and for delivering health benefits from studies made on other plants, to jump to the conclusion that consuming mangoes actually supplies such benefits to real human consumers. For that, studies performed on cell models and animal models, or, even much better, clinical studies are required. Table 3 summarizes the positive health effects of mango as they may be derived from studies on cells and animals, and from clinical studies.

Table 3 Potential positive health effect of mango (*Mangifera indica* L.) evaluated in clinical studies, mammalian cells or animal models

Mangoes (pulp/juice/whole fruit or extract)	Type of trial	Study details	Major experimental findings	References
Mango pulp ('Keitt')	Clinical study	11 healthy volunteers (21–38 years) consumed 400 g day^{-1} of pulp during 10 days. Metabolites of gallic acid (GA) in urine excreted over a 12 h period were quantified	Seven metabolites of GA were identified in the urine of healthy volunteers, and two microbial metabolites were found to be significantly more excreted	Barnes et al. (2016)
Whole and fresh cut mango ('Ataulfo')	Clinical study	During 30 consecutive days, 30 normolipidaemic volunteers (20–50 years) received daily 200 g of whole mango or fresh cut mango	Hypertriglyceridemia was prevented	Robles-Sánchez et al. (2011)
Mango (juice, pulp and dried fruit)	Clinical study	Healthy volunteers (24–25 years) were served breakfast daily, which included bread, yogurt and one of the three forms of mango fruit (juice, fresh and dry slice). Blood samples were collected three times: during fasting, 4 and 8 h after the test meals	The diet contributed efficiently to improve the vitamin A status	Gouado et al. (2007)
Mango (purified mangiferin)	Human umbilical vein endothelial cells (HUVEC)	HUVEC were treated with different concentrations of mangiferin (10 µg mL^{-1} or 20 µg mL^{-1}), and incubated at 37°C in 5% CO_2 atmosphere for 24 h	Purified mangiferin showed protective effect on H_2O_2-treated HUVEC in a dose-responsive manner	Luo et al. (2012)
Mango extracts ('Ataulfo' and 'Haden')	Human SW-480 colon cancer cells	Cells were treated with mango polyphenolics (5 and 10 mg of GAE L^{-1}) for 24 h and harvested for flow cytometer analysis	Polyphenolics exerted protection of normal colon cells (CCD-18Co) by lowering the ROS generation in a dose-dependent manner	Norato et al. (2010)

(Continued)

Table 3 (Continued)

Mangoes (pulp/juice/ whole fruit or extract)	Type of trial	Study details	Major experimental findings	References
Mango extracts ('Irwin')	Human hepatoma cell line (HepG2)	Cell lines were incubated with mango extracts for 24 h to allow cell attachment before exposure to varying concentrations of mango polyphenolics	Peel extract exhibited significant anti-proliferative and antioxidant effect against all tested cancer cell lines when compared to pulp in a dose-dependent manner	Kim et al. (2010)
Mango juice ('Ubá')	Male Wistar rats ($n = 32$)	The biometry and biochemical parameters were evaluated in four experimental groups. Peroxisome proliferator-activated receptor gamma (PPAR-g), lipoprotein lipase and fatty acid synthase expression, tumour necrosis factor-a and interleukin-10 (IL-10), as well as histomorphology of the epididymal adipose tissue were determined	Mango juice showed modulatory effects on both inflammation and adipogenesis, indicating potential to prevent and combat obesity	Natal et al. (2016)
Mango (peel and pulp)	Male Wistar rats ($n = 20$)	The experimental groups were fed with either mango pulp or mango peel. An hour after feeding, 150 µL of 50% ethanol was administered orally. An hour after the ethanol was given, blood was drawn from the heart	Peel and pulp decreased the mouse plasma ethanol levels and increased the activities of alcohol dehydrogenase and acetaldehyde dehydrogenase activities	Kim et al. (2011)
Mango pulp ('Ataulfo')	Female rats ($n = 116$)	Mango was administered in the drinking water (0.02–0.06 g mL⁻¹) during both short-term and long-term (LT) periods to rats treated or not with N-methyl-N-nitrosourea (MNU). The plasma antioxidant capacity was evaluated by the FRAP method	The plasma antioxidant capacity (FRAP assay) tended to increase in a dose-dependent manner in the LT rats not treated with MNU	García-Solís et al. (2008)

2.4 Cell studies

Luo et al. (2012) investigated the beneficial effects of purified mangiferin on human umbilical vein endothelial cells (HUVEC) under H_2O_2-induced stress. Luo et al. (2012) concluded that mangiferin at a concentration of either 10 or 20 µg mL^{-1} showed substantial protective effects on HUVEC with survival rates substantially improved at 0.0625 mmol H_2O_2 L^{-1}, 0.125 mmol H_2O_2 L^{-1} and 0.25 mmol H_2O_2 L^{-1}. Their observations show that purified mangiferin exerts protective effects against oxidative stress, but the underlying mechanism remains unclear requiring further investigations.

Norato et al. (2010) evaluated the anti-carcinogenic effects of polyphenols from different mango varieties in human SW-480 colon cancer cells and non-cancer CCD-18Co colon cells. Polyphenolics present in 'Ataulfo' and 'Haden' inhibited the growth of SW-480 colon cancer cells. SW-480 gene regulation included induction of apoptosis in addition to cell cycle arrest in the G2/M phase. 'Ataulfo' and 'Haden' polyphenolics exerted also protection of non-cancer CCD-18Co colon cells by lowering the ROS generation in a dose-dependent manner.

Kim et al. (2010) evaluated the antioxidant and anti-proliferative properties of pulp and peel mango in a human hepatoma cell line (HepG2). They observed that peel extract has anti-proliferative and antioxidant effects in all tested cancer cell lines in a dose-dependent manner. They moreover observed that the effects were correlated with phenolic and flavonoid contents.

2.5 Studies on animals

Recently, Natal et al. (2016) studied the effect of 'Ubá' mango juice on adiposity and inflammation in male obese Wistar rats after a high-fat diet. Their findings show that mango juice improves the gene expression related to adiposity and inflammation, and also decreased several biochemical, cytological and biometrical markers in obese rats to levels similar to the ones found in control rats. Natal et al. (2016) concluded that 'Ubá' mangoes have potential as a functional food to prevent and combat obesity.

Kim et al. (2011) investigated the ameliorating effect of mangoes on plasma ethanol levels using a mouse model. ^{1}H-NMR (spectroscopy) was employed to investigate the differences in metabolic profiles of mango fruits, and mouse plasma samples fed with mango fruit. Results confirm that mango samples (peel and pulp) remarkably decreased the mouse plasma ethanol levels and increased the activities of alcohol dehydrogenase and acetaldehyde dehydrogenase, reducing hangover symptoms in treated animals.

2.6 Clinical studies

Gouado et al. (2007) evaluated the bioavailability of carotenoids (α- and β-carotene and lycopene) present in mango and papaya consumed in three forms (juice, fresh and dried). Two groups of seven healthy volunteers, each were submitted to three types of meal treatments (juice, fresh and dried fruit). All the treatments lasted only one day during which blood samples were collected three times: during fasting (T0), 4 h and 8 h after the test meal. A comparison between the three forms revealed that papaya and mangoes consumed in the form of juice or fresh fruit lead to the highest bioavailability values (Gouado et al., 2007).

Robles-Sánchez et al. (2011) evaluated the influence of intake of 'Ataulfo' mangoes, either whole or under the form of fresh cuts, on lipids and antioxidant capacity of healthy adults' plasma. Thirty normolipidaemic volunteers were randomly divided into two groups (whole mangoes and fresh cuts). During 30 consecutive days these volunteers received daily 200 g of whole mangoes or fresh cuts. Lipid levels and antioxidant capacity in plasma were determined at the onset of the trial, as well as 15 and 30 days after. Serum triglycerides were reduced by 37% and 38%, respectively, after 30 days of supplementation with whole mangoes and fresh cuts. Very low-density lipoprotein cholesterol levels were reduced in a similar proportion. Both treatments increased plasma antioxidant capacity measured by ORAC and TEAC methods. The authors suggested that addition of mango fruit to generally accept healthy diets could have a beneficial effect preventing hypertriglyceridaemia, and that fresh cut processing does not affect the beneficial properties of mango.

Barnes et al. (2016) conducted a clinical study pilot with healthy volunteers that consumed 400 g day^{-1} of mango pulp (cv. Keitt) for 10 days. They characterized and quantified seven metabolites of GA in urine excreted over a 12 h period. A significant increase in the excretion of pyrogallol-O-sulphate and deoxypyrogallol-O-sulphate were observed between days 1 and 10, increasing from 28.5 to 55.4 mg L^{-1} and from 23.6 to 47.7 mg L^{-1}, respectively. Additionally, the *in vitro* hydrolysis of GTs was monitored at physiological pH and temperature conditions. After 4 h a shift in composition from relativity high to low molecular weight GTs was observed. Seven metabolites of GT were identified in the urine of healthy volunteers, and two microbial metabolites were found to be excreted in excess following 10 days of mango consumption. Mango GTs were also found to release free GA in conditions similar to the intestines. GTs may serve as a pool of pro-GA compounds that can be absorbed or can undergo microbial metabolism (Barnes et al., 2016).

3 Increasing phytochemical concentrations in mango fruits

Enough evidence has been accumulated through epidemiologic and clinical studies, first about the global benefits of FAVs in the human diet and second about the dietary effects of the phytochemicals they supply, especially vitamins and secondary metabolites. On the basis of such undisputed evidence, even in the absence of precise recommendations, it makes sense to encourage people to consume more FAVs. Unfortunately, the five-a-day campaigns in developed countries to persuade people to eat at least five portions of FAVs every day have proven to be a relative failure so far, and the situation is no better in developing countries. Taking these facts into account, it appears reasonable trying to improve the current situation by encouraging, besides the consumption of FAVs, the consumption of foods and food supplements with enhanced concentrations in phytochemicals. Within this view, the proposition to produce mangoes with increased concentrations in phytochemicals can be considered (Poiroux-Gonord et al., 2010).

We shall review now successively the effects of genetic factors and of the ripening stage, before considering the environmental levers that can be used before and after harvest.

3.1 Genetic factors and fruit-to-fruit variability (see Table 1)

Manthey and Perkins-Veazie (2009) compared five varieties of mango ('Ataulfo', 'Haden', Keitt', 'Kent' 'and 'Tommy Atkins') from four countries. They observed that vitamin C ranged from 11 to 134 mg 100 g^{-1} of pulp puree, and that β-carotene varied from 5 to 30 mg kg^{-1} among the five varieties. Total phenolic content ranged from 19.5 to 166.7 mg of GA equivalents (GAEs) 100 g^{-1} of puree. The varieties 'Haden', 'Keitt', 'Kent' and 'Tommy Atkins' had similar total phenolic contents with an average of 31.2 ± 7.8 mg GAE 100 g^{-1} of puree, whereas the 'Ataulfo' variety contained substantially higher amounts. In contrast, the country of origin and harvest dates had far less influence on these parameters. 'Ataulfo' mangoes contained significantly higher amounts of mangiferin and ellagic acid than the other four varieties. Large fruit-to-fruit variations in the concentrations of these compounds were observed within sets of mangoes of the same cultivar with the same harvest location and date. Since phytochemicals are often endowed with strong antioxidant properties, the large differences observed in concentrations in phytochemicals as a consequence of genetic factors should be reflected in large differences in total antioxidant capacity. Shi et al. (2015) evaluated 28 mango cultivars for phytochemical content and antioxidant capacity. Using a total antioxidant potency composite index, they found that antioxidant capacity varied indeed strongly among all the genotypes, from 6.12 to 81.39.

   While the strong effects of genetic factors tend to overshadow other effects, fruit-to-fruit variability points again towards the importance of pre-harvest factors. Indeed the within-tree variability and the tree-to-tree variability arguably originate from differences in water availability among trees, differences in photosynthesis and carbon gains, as well as differences in photo-oxidative stress, as a consequence of differences in fruit position in the canopy and of drought-associated differences in stomatal conductance and sugar accumulation in leaves close to fruits. Moreover differences in carbon status and in oxidative stress necessarily translate into differences in ripening stages which again have a tremendous influence on concentrations in phytochemicals of fruits.

3.2 Effect of the harvesting stage

According to Kondo et al. (2005), contents in total phenolics and in vitamin C of the mango pulps increase and decrease, respectively, during fruit growth, from 14 to 56 days after full bloom (DAFB). Contents then appear relatively stable from 56 to 84 DAFB. But there are contradictory observations. Wenkam (1979) observed a threefold decrease in vitamin C content of 'Haden' fruits from green to ripe. Kim et al. (2007) observed similarly that phenolic compounds decrease as fruit ripens. More specifically, GA and, in general, total hydrolysable tannins were found to decrease by 22% and 57%, respectively (Kim et al., 2007).

   The picture is different when considering carotenoids. Observations on 'Cogshall' fruits indicate that it is advantageous for the carotenoid content to delay the harvesting time (Joas et al., 2012). The majority of carotenoids in mango fruit are isomers of violaxanthin and β-carotene, whose contents will effectively increase in relation to maturity (Mercadante and Rodriguez-Amaya, 1998). The highest content in carotenoids is achieved for fruits well-exposed to sunlight, harvested at the latest maturity stage, and allowed to ripe at 20°C.

Since the harvesting stage has such a high impact on the phytochemical content of mango fruits, it seems desirable to acquire parameters to assess harvesting maturity, more relevant and precise than the usual criteria, such as colour changes. Chlorophyll fluorescence has been proposed as a credible tool to reduce biases associated with growth conditions (Lechaudel et al., 2010).

Maciel et al. (2010) evaluated the potential for cultural practices to influence contents in total phenolics and flavonoids, as well as antioxidant activity of mangoes by comparing at three maturation stages of 'Tommy Atkins' fruits from biodynamic culture, organic farming and conventional growing. They observed the highest antioxidant activity in mature green and ripe fruits from biodynamic culture, and the highest antioxidant activity in unripe fruits from organic farming, which they related to the flavonoid content of fruits. Interestingly, the observations of Maciel et al. (2010) are consistent with observations made by Oliveira et al. (2013) of positive effects of organic farming on contents in vitamin C and phenolics of tomato fruits.

It was observed under conventional farming that the highest carotenoid content is obtained in mangoes from girdled branches at 25 leaves per fruit (high fruit load – low carbon supply to fruits) rather than in fruits from girdled branches at 100 leaves per fruit (low fruit load – high supply of carbon to fruits) (Table 4) (Joas et al., 2012). This observation does not support the common view that carotenoid biosynthesis competes for carbohydrate supply with structural growth or sugar build-up, and thus contrasts with other reports showing that carbohydrate limitation negatively impacts carotenoids. Poiroux-Gonord et al. (2012a; 2012b) suggested that early changes in carbohydrate availability in fruiting branches of *Citrus* could provide a signal for carotenoid regulation, independent of the substrate availability *per se*. More specifically, early carbon limitation may increase the potential for fruit carotenoid synthesis and accumulation, and enhance final carotenoid content through enhancing plastid storage capacity (Poiroux-Gonord et al., 2012a; Poiroux-Gonord et al., 2012b). So far, such observations have not been made in mango even though the observations of Joas et al. (2012) suggest that such a mechanism may well be at play in mango.

Exposure to light may also influence carotenoid content of mangoes. There are substantial differences in carotenoid content between shaded and well-exposed mango fruits (Joas et al., 2012), and in vitamin C between the shaded and the well-exposed sides of fruits (Léchaudel et al., 2013). Clearly, training and pruning practices which aim at increasing light penetration to fruits could be used as a lever to increase concentrations in carotenoids and in vitamin C of mango fruits.

Table 4 Major findings about the effects of pre-harvest factors influencing the concentrations in phytochemicals of the pulp of mangoes at the time of harvest

Technique used	Cv.	Major effects observed (expressed on a fresh weight basis)	Reference
Leaf-to-fruit ratio (25 vs. 100)	'Cogshall'	+67% carotenoids when fruits are allowed to ripe at 20°C after harvest	Joas et al. (2012)
Light exposition of fruits (well exposed vs. shaded)	'Cogshall'	Global positive trend for carotenoids	Joas et al. (2012)

3.3 Stress as a lever to increase phytochemical content

Global climate change entails many threats and challenges for the majority of crops. Even though increasing temperatures stimulate mango photosynthesis up to ca. 40°C (Schaffer et al., 2009), a reduction in yield must be expected as a consequence of the stressing conditions associated with the climate change (Normand et al., 2015). Fruit crops will certainly suffer from the increased extension of drought conditions among others; however, yield is arguably not as important for fruit as for grain crops or oil crops. Yield does matter for fruit crops, but quality criteria are as important if not more important (Ripoll et al., 2014). Fruits are expected to supply health benefits and to bring hedonistic pleasures associated with specific aromatic compounds. We may thus distance ourselves from the dominant deleterious effect of stress on crop performance and consider the potential benefits (Ripoll et al., 2014). Fruits from stressed trees may in particular display a higher content in health-promoting phytochemicals.

Here is a brief summary of the physiological mechanisms behind the idea that stress may be beneficial when it comes to synthesis and accumulation of phytochemicals in fruits. For more details, see the following articles: Fanciullino et al. (2013), Poiroux-Gonord et al. (2013) and Ripoll et al. (2014).

Photosynthesis represents the major source of ROS in green plants (Asada, 1999). The mechanism is now well understood. Most stresses that result in the inhibition of the Calvin cycle, may result in an excess of energy entering the system under the form of photons when compared to the quantity of energy used by photochemistry, and a decrease in reoxidation of NADPH. This leads to the formation of singlet oxygen $^1O_2$ at the level of photosystem II, PSII (Apel and Hirt, 2004) and to the transfer of electrons to molecular oxygen, that is, the formation of superoxide O_{2-} at photosystem I (PSI). Plants are well prepared to cope with such conditions. Non-photochemical quenching allows for the dissipation of excess excitation energy in the light-collecting antennae of PSI, while the photosynthetic electron transport rate of stressed plants is reallocated from photosynthesis to photorespiration and the Mehler reaction at PSI. The glycolate oxidase and the Mehler peroxidase reactions lead to the production of substantial amounts of H_2O_2 (Noctor et al., 2002; Smirnoff, 1993) (a less reactive ROS than $^1O_2$ and O_{2-}), in peroxisomes and chloroplasts, respectively. Besides catalase (CAT) there are several enzymes and enzymatic systems to eliminate H_2O_2.

Ultraviolet light, like other stresses, may be at the origin of photo-oxidative stress, that is, the production of ROS associated with the functioning of the photosynthetic machinery (Urban et al., 2016). Similarly, drought is at the origin of photo-oxidative stress or exacerbates it by reducing stomatal conductance, the amount of CO_2 feeding the Calvin cycle, and therefore, reoxidation of NADPH (Grassmann et al., 2002). Eventually, a decrease in translocation of sugars synthetized in leaves inhibits the Calvin cycle and reduces reoxidation of NADPH.

According to Fanciullino et al. (2013) and Poiroux-Gonord et al. (2013), oxidative stress can originate both in the fruit and in the leaf, from which it propagates to nearby fruit through an unknown signal. Activation of redox-sensitive systems upregulates the transcription of genes involved in biosynthetic pathways, leading to higher levels of corresponding proteins and higher levels of secondary metabolites. Redox changes also regulate the activity of the enzymes of the biosynthetic pathways. The increase in ROS can stimulate production of secondary metabolites indirectly by promoting fruit development and, as far as carotenoids are concerned, conversion of chloroplasts into chromoplasts.

Besides regulated deficit irrigation, among stressing conditions that could be used in mango trees, girdling, a common practice among mango growers, could be a powerful technique to force accumulation of sugars in leaves by suppressing phloem connections and therefore provoke photo-oxidative stress in leaves (Urban and Léchaudel, 2005; Urban and Alphonsout, 2007) and the subsequent accumulation of phytochemicals in fruits.

4 Pre- and post-harvest factors influencing bioactive compounds of mango fruits

4.1 Ripening after harvest

Ripening climacteric fruits like mangoes, even when they are detached from the parent plant, undergo physiological, biochemical and molecular changes that directly affect their quality traits (Osorio and Fernie, 2013; Prasanna et al., 2007).

Ibarra-Garza et al. (2015) evaluated the effects of post-harvest ripening (6 ripening stages, RS) on the nutraceutical and physico-chemical properties of 'Keitt' mangoes. Based on measurements of antioxidant activity and of vitamin C, total phenolics and carotenoids they found that the optimum ripening stages were RS2 and RS6. Razzaq et al. (2013) evaluated the impact of ripening duration on the antioxidant capacity of 'Samar Bahisht Chaunsa' mangoes. They observed an increase in dismutase superoxide (SOD) activity during ripening and found that total antioxidant activity peaked at day 7. Palafox-Carlos et al. (2012) observed that antioxidant capacity of 'Ataulfo' mangoes was tightly associated with contents in phenolics and flavonoids, and increased from RS2 (20 to 30% yellow colour surface) to RS3 (70–80%).

4.2 Post-harvest handling

Post-harvest handling, processing and storage conditions have the potential to strongly influence the content in phytochemicals of mango fruits as well as their antioxidant potential. They may help to maintain positive characteristics and even boost them.

Khaliq et al. (2016) observed that gum arabic (GA) (10% w/v) coating enriched with calcium chloride (CA) (3% w/v) maintained high DPPH radical scavenging activity of mango fruits (cv Chok Anan) stored at low temperature (Table 2). They found, moreover, that GA either alone or in combination with CA effectively inhibited the loss of phenolics and ascorbic acid.

Junmatong et al. (2015) studied the effect of long-term cold storage (5°C) combined with salicylic acid (SA) at 1 mM on antioxidants of 'Nam Dok Mai' mangoes. They observed that SA-treated mango fruits exhibited significantly higher levels of vitamin C, total phenolics and antioxidant activity.

In another study, Kaushik et al. (2014) examined the effect of high-pressure processing (100 to 600 MPa for 1 s to 20 min) on colour, biochemical and microbiological characteristics of mango pulp (cv. Amrapali). They observed a retention of 92 % of total phenolics and of 90 % for antioxidant activity in treated mangoes.

The impact of processing of fresh mangoes was evaluated by (Gil et al., 2006) by comparing fresh cut and whole fruits, and found that fresh cutting resulted in less than 5% loss in vitamin C, 10 to 15% loss in carotenoids and no significant loss in total phenolics after 6 days at 5°C. In another study, Robles-Sánchez et al. (2009) demonstrated that dipping cubes of 'Kent' mangoes in ascorbic acid (1%) increased antioxidant activity.

The most spectacular effects of phytochemicals were observed with light. Gil et al. (2006) observed strong positive effects of visible light on carotenoids content of 'Ataulfo' mangoes. Jiang et al. (2015) evaluated the effect of exposure of 'Tainung' mangoes to ultraviolet B (UV-B) radiation at 5 kJ m^{-2} delivered over 4 h. They found that the UV-B associated strong increase in antioxidant compounds (vitamin C and phenolics) was highly correlated to reduced ROS level (H_2O_2 and O_2_) and to increased activities of SOD and CAT. While exposure to UV-C light or to infrared (IR) radiation may lead to contrasting effects according to the considered phytochemicals, PL which encompasses UV-C, UV-B, UV-A, visible and near-IR radiations was found to be at the origin of strong positive effects for vitamin C, phenolics and carotenoids as well. Apparently the boosting effect of PL on phytochemical content of mango fruits can be observed on fresh cuts (Charles et al., 2013) and on entire fruits (Lopes et al., 2015). We attached to this review a case study about the effect of PL on phytochemical content of 'Tommy Atkins' entire mangoes (see below).

5 Case study: low fluence PL to enhance mango phytochemical content

The phenomenon known as hormesis refers to physiological stimulation of beneficial responses by low levels of stressors, thus in theory, it may be used as a promising tool in the food industry leading to healthier products by enhancing phytochemical levels of either whole or fresh cut produce (González-Aguilar et al., 2010c; Bravo et al., 2012). However, little is known about the physiological basis of the accumulation of phytochemicals as a response to a post-harvest stress. In a specific case of study, Lopes et al. (2015) tested the hypothesis that a hormetic dose of PL (100–1100 nm) was capable to induce biochemical changes in the tissues (peel and pulp) of 'Tommy Atkins' mangoes.

Physiologically mature 'Tommy Atkins' mangoes were submitted to PL at a fluence of 0.6 J.cm^{-2} (2 pulses of 0.3 J.cm^{-2} each) and then, stored for 7 days at 20°C. Fruit pulp and peel were separated and evaluated for H_2O_2 content (Sergiev et al., 2001) as an indicator of oxidative stress, vitamin C content (Strohecker and Henning, 1967), SOD (Giannopolitis and Ries, 1977), CAT (Beers and Sizer, 1952) and ascorbate peroxidase (Nakano and Asada, 1981) activities as indicators of antioxidant metabolism. Total carotenoids were measured as described by Lichtenthaler and Buschmann (2001), and the results were expressed as mg kg^{-1}. Polyphenolic pigments as total anthocyanins and yellow flavonoids were evaluated as described by Francis (1982), with results expressed as mg kg^{-1}. Total phenol content of mangoes was measured by a colorimetric assay using Folin–Ciocalteu reagent as described by Larrauri et al. (1997) and Obanda et al. (1997) and expressed as GAE mg kg^{-1}. Phenylalanine ammonia lyase (PAL) activity was assayed as described by Mori et al. (2001) and El-Shora (2002), with slight modifications. PAL-specific activity was expressed as µmol *trans*-cinnamic acid h^{-1} mg^{-1} P.

After 7 days of storage, the PL treatment made at the onset of the storage period had resulted in a +350% increase in total carotenoid content in the pulp, when compared to the control. In the peel, the PL had similarly resulted in a +90% increase in carotenoid content when compared to the control. Storage time and PL did not influence the anthocyanin and yellow flavonoid contents in the peel whereas PL resulted in a +21% increase in anthocyanins in the pulp and, similarly, a +42% increase in yellow flavonoid after 7 days of storage. PL also strongly helped to reduce the storage-associated decrease in vitamin C content in the pulp (Table 5).

Table 5 Major findings about the effects of post-harvest techniques on the concentrations in phytochemicals of the pulp of mangoes either under the form of entire fruits or under the form of fresh cuts

Technique used	Entire fruits/ fresh cuts (cv)	Major effects observed (expressed on a fresh weight basis if not specified otherwise)	References
High pressure (100 to 600 Mpa, 1 s to 20 min)	Pulp ('Amrapali')	+85% vitamin C +92% total phenolics	Kaushik et al. (2014)
Electron beam ionizing radiation (3.1 kGy)	Entire fruits ('Tommy Atkins')	−54% vitamin C −74% carotenoids (vs. control after 18 days storage)	Reyes and Cisneros-Zevallos (2007)
High electric field (150 kV/m of electric field for 45 min)	Entire fruits ('Irwin')	Decrease in β-carotene, no clear effect on total phenolics, quercetin and vitamin C (vs. control after 20 days storage)	Shivashankara et al. (2004)
Light exposure (4 to 5 μmol photons m^{-2} s^{-1})	Fresh cuts ('Ataulfo')	+228% total carotenoids	Gil et al. (2006)
Low fluence pulsed light (2 pulses amounting to 0.6 J cm^{-2})	Entire fruits ('Tommy Atkins')	+60% vitamin C +30% total phenolics +350% total carotenoids (vs. control after 7 days storage)	Lopes et al. (2015)
Pulsed light (2 pulses amounting to 8 J cm^{-2})	Fresh cuts ('Kent')	+55% vitamin C (nmol g^{-1} DM) +350% total carotenoids (vs. control after 7 days storage)	Charles et al. (2013)
UV-B light (0.5 J cm^{-2} delivered over 4 hours)	Entire fruits ('Tainung', cold-stored)	Strong positive effect on vitamin C, quercetin, kaemferol and gallic acid	Jiang et al. (2015)
UV-C light (2.46 and 4.93 J cm^{-2})	Entire fruits ('Haden')	+14 to 25% total phenolics +20 to 80% total flavonoids (vs. control after 15 days storage)	Gonzalez-Aguilar et al. (2007a)
UV-C light (1 to 10 mn, using 15 W lamps at a 15 cm distance)	Fresh cuts ('Tommy Atkins')	Down to −71% vitamin C down to −67% β-carotene up to +28% total phenolics up to +50% total flavonoids (vs. control after 15 days storage)	Gonzalez-Aguilar et al. (2007b)

(Continued)

Table 5 *(Continued)*

Technique used	Entire fruits/ fresh cuts (cv)	Major effects observed (expressed on a fresh weight basis if not specified otherwise)	References
UV-C light (4.93 J cm^{-2})	Entire fruits ('Nam Dok Mai')	Negative effect on total phenolics (on a DM basis) (vs. control after 15 days storage)	Safitri et al. (2015)
IR treatment (5, 10 and 15 min)	Fresh cuts ('Tommy Atkins')	Down to −80% vitamin C contrasted effects on carotenoids up to +247% total phenolics (vs. control after 16 days storage)	Sogi et al. (2012)
Osmo-dehydrofreezing (sucrose, glucose and maltose at 45%)	Pulp ('Tainung')	Increase in phenolics like p–hydroxybenzoic acid, quercetin, p-coumaric acid and sinapic acid	Gil et al. (2006)
Hot water immersion at 50°C for 30 min followed by cooling for 15 min	Fresh cuts ('Tommy Atkins)	Slight positive but NS effect on total carotenoids	Djioua et al. (2009)
Cold shock (0°C for 4 h)	Entire fruits ('Wacheng')	Slight positive but NS effect on vitamin C +70% total phenolics (vs. control after 12 days storage)	Zhao et al. (2006)
Hot water immersion at 46.1°C for 70 to 110 min	Entire fruits ('Tommy Atkins')	Negative effect on phenolics (vs. control after 4 days storage)	Kim et al. (2009)
Hot water immersion at 50°C for 60 min	Entire fruits ('Tommy Atkins')	No significant difference in total carotenoids (vs. control after 16 days storage at 5°C)	Talcott et al. (2005)
Salicylic acid (2 mM for 5 min)	Entire fruits ('Chausa')	+27% total phenolics NS effect on carotenoids (vs. control after 30 days)	Barman and Asrey (2014)
Salicylic acid or oxalic acid (2 and 5 mM for 10 min)	Entire fruits ('Zill')	Higher % of vitamin C under the reduced form (vs. control after 30 days storage at 5°C)	Ding et al. (2007)

H_2O_2 content was evaluated as an indicator of oxidative stress, and found to increase in response to PL, but only in the pulp. SOD activity increased by 58% in both tissues of PL-treated mangoes, when compared to the control. CAT activity was also enhanced

(+104%) by PL, but only in the pulp. The concomitant increase in H_2O_2 levels and activities of antioxidant enzymes provides evidence that PL was at the origin of oxidative stress, that is, the production of ROS, and the subsequent triggering of antioxidant mechanisms (Jaleel et al., 2009).

Moreover, PAL activity was significantly enhanced by PL treatment in both mango tissues which is consistent with the higher total phenolic content found in PL-treated mango tissues. PAL has been commonly used as an indicator of stressful conditions (Sreelakshmi and Sharma, 2008) and together with higher H_2O_2 levels and activities of the antioxidant enzymes, the increase in PAL activity suggests that PL induced an oxidative imbalance in cells of mango pulp and peel.

In conclusion, it may be stated that PL positively affected the post-harvest physiology of 'Tommy Atkins' mangoes by strongly stimulating synthesis and accumulation of health-promoting phytochemicals, namely carotenoids and phenolics, while limiting the decrease in vitamin C normally observed during storage. PL emerges from our trial as an exceptionally potent technique for improving quality of mangoes that clearly deserves more attention from the scientific community in the future.

6 Future trends and conclusion

Mango fruits represent an outstanding source of phytochemicals. There is, however, a large variability in contents and therefore quality, due to the influence of numerous genetic factors and environmental factors before harvest, not to mention the roles of harvesting time, and of ripening and storage conditions and duration. Control of quality is of paramount importance to the mango industry. It is also a major health issue for consumers and stakeholders. There is therefore the need both to develop a better understanding of the way factors and their interactions influence synthesis and accumulation of phytochemicals in mangoes, and to develop innovative tools and techniques, endowed with a potent ability to drive the secondary metabolism in the fruits. We specifically advocate here for more studies to be conducted about light, especially about PL after harvest on both entire fruits and fresh cuts. In addition to more observations about doses and possibly repetitions of doses, what we need now is to go further, beyond descriptive studies, for acquiring insight in the physiological mechanisms involved in light sensing and signalling in relationship with the metabolic pathways of ascorbate, phenolic compounds and carotenoids.

7 Where to look for further information

For those who are interested in developing an integrated view of the way environmental factors and their interactions influence synthesis and accumulation of phytochemicals in fruits, we recommend to read the article of Fanciullino et al. (2013). The article is focusing on carotenoids but most of the rationales developed in this review paper also apply to phenolic compounds and, to a lesser extent, to vitamin C. We also strongly advise readers to increase their awareness of the existing debates about the concept of antioxidant and about the mechanisms of action of phytochemicals after ingestion by human consumers. The articles of Halliwell (2013) and of Anhê et al. (2015) are must-reads in our view.

8 References

Ames, B. N., Shigenaga, M. K. and Hagen, T. M. (1993). Oxidants, antioxidants, and the degenerative diseases of aging. *Proceedings of the National Academy of Sciences of the United States of America* 90:7915–22.

Anhê, F. F., Roy, D., Pilon, G., Dudonné, S., Matamoros, S., Varin, T. V., Garofalo, C., Moine, Q., Desjardins, Y. and Levy, E. (2015). A polyphenol-rich cranberry extract protects from diet-induced obesity, insulin resistance and intestinal inflammation in association with increased Akkermansia spp. population in the gut microbiota of mice. *Gut* 64:872–83.

Apel, K. and Hirt, H. (2004). Reactive oxygen species: Metabolism, oxidative stress, and signal transduction. *Annual Review of Plant Biology* 55:373–99.

Arrigoni, O. and De Tullio, M. C. (2002). Ascorbic acid: much more than just an antioxidant. *Biochimica et Biophysica Acta (BBA) – General Subjects* 1569:1–9.

Asada, K. (1999). The water-water cycle in chloroplasts: Scavenging of active oxygens and dissipation of excess photons. *Annual Review of Plant Physiology and Plant Molecular Biology* 50:601–39.

Baltschun, D., Beutner, S., Briviba, K., Martin, H. D., Paust, J., Peters, M., Rover, S., Sies, H., Stahl, W., Steigel, A. and Stenhorst, F. (1997). Singlet oxygen quenching abilities of carotenoids. *Liebigs Annalen-Recueil* 1887–93.

Barman, K. and Asrey, R. (2014). Salicylic acid pre-treatment alleviates chilling injury, preserves bioactive compounds and enhances shelf life of mango fruit during cold storage. *Journal of Scientific & Industrial Research* 73:713–18.

Barnes, R. C., Krenek, K. A., Meibohm, B., Mertens-Talcott, S. U. and Talcott, S. T. (2016). Urinary metabolites from mango (*Mangifera indica* L. cv. Keitt) galloyl derivatives and in vitro hydrolysis of gallotannins in physiological conditions. *Molecular Nutrition & Food Research* 60(3): 542–50. DOI:10.1002/mnfr.201500706.

Ben-Amotz, A. and Fishier, R. (1998). Analysis of carotenoids with emphasis on 9-cis β-carotene in vegetables and fruits commonly consumed in Israel. *Food Chemistry* 62:515–20.

Berardini, N., Fezer, R., Conrad, J., Beifuss, U., Carle, R. and Schieber, A. (2005). Screening of mango cultivars for their contents of flavanol O and xanthone C-glycosides, anthocyanins and pectins. *J Agric. Food Chem* 52:1563–70.

Beutner, S., Bloedorn, B., Frixel, S., Blanco, I. H., Hoffmann, T., Martin, H. D., Mayer, B., Noack, P., Ruck, C., Schmidt, M., Schulke, I., Sell, S., Ernst, H., Haremza, S., Seybold, G., Sies, H., Stahl, W. and Walsh, R. (2001). Quantitative assessment of antioxidant properties of natural colorants and phytochemicals: carotenoids, flavonoids, phenols and indigoids. The role of β-carotene in antioxidant functions. *Journal of the Science of Food and Agriculture* 81:559–68.

Charles, F., Vidal, V., Olive, F., Filgueiras, H. and Sallanon, H. (2013). Pulsed light treatment as new method to maintain physical and nutritional quality of fresh-cut mangoes. *Innovative Food Science & Emerging Technologies* 18:190–5.

Chen, A. Y. and Chen, Y. C. (2013). A review of the dietary flavonoid, kaempferol on human health and cancer chemoprevention. *Food Chemistry* 138:2099–107.

Chen, J., Tai, C. and Chen, B. (2004). Improved liquid chromatographic method for determination of carotenoids in Taiwanese mango (*Mangifera indica* L.). *Journal of Chromatography A* 1054:261–8.

Combs, G. F. (1995). *The Vitamins: Fundamental Aspects in Nutrition and Health*. Academic Press, San Diego.

Corral-Aguayo, R. D., Yahia, E. M., Carrillo-Lopez, A. and Gonzalez-Aguilar, G. (2008). Correlation between some nutritional components and the total antioxidant capacity measured with six different assays in eight horticultural crops. *Journal of Agricultural and Food Chemistry* 56:10498–504.

Das, L., Bhaumik, E., Raychaudhuri, U. and Chakraborty, R. (2012). Role of nutraceuticals in human health. *Journal of Food Science and Technology* 49(2):173–83. DOI:http://dx.doi.org/10.1007/s13197-011-0269-4. PMid:23572839.

Devasagayam, T. P. A., Steenken, S., Obendorf, M. S. W., Schulz, W. A. and Sies, H. (1991). Formation of 8-hydroxy(deoxy)guanosine and generation of a strand breaks at guanine residues in DNA by singlet oxygen. *Biochemistry* 30:6283–9.

Ding, Z.-S., Tian, S.-P., Zheng, X.-L., Zhou, Z.-W. and Xu, Y. (2007). Responses of reactive oxygen metabolism and quality in mango fruit to exogenous oxalic acid or salicylic acid under chilling temperature stress. *Physiologia Plantarum* 130:112–21.

Djioua, T., Charles, F., Lopez-Lauri, F., Filgueiras, H., Coudret, A., Freire, Jr. M., Ducamp-Collin, M.-N. and Sallanon, H. (2009). Improving the storage of minimally processed mangoes (*Mangifera indica* L.) by hot water treatments. *Postharvest Biology and Technology* 52:221–6.

Fanciullino, A. L., Bidel, L. P. R. and Urban, L. (2013). Carotenoid responses to environmental stimuli: integrating redox and carbon controls into a fruit model. *Plant, Cell & Environment*:n/a-n/a. DOI: 10.1111/pce.12153.

Francis, F. (1982). Analysis of anthocyanins. In P. Markakis (ed.), *Anthocyanins as food colors*, pp. 181–207.

Franke, A. A., Custer, L. J., Arakaki, C. and Murphy, S. P. (2004). Vitamin C and flavonoid levels of fruits and vegetables consumed in Hawaii. *Journal of Food Composition and Analysis* 17:1–35.

García-Solis, P., Yahia, E. M. and Aceves, C. (2008). Study of the effect of 'Ataulfo' mango (*Mangifera indica* L.) intake on mammary carcinogenesis and antioxidant capacity in plasma of N-methyl-N-nitrosourea (MNU)-treated rats. *Food Chemistry* 111: 309–15.

Gil, M. I., Aguayo, E. and Kader, A. A. (2006). Quality changes and nutrient retention in fresh-cut versus whole fruits during storage. *Journal of Agricultural and Food Chemistry* 54:4284–96.

Gonzalez-Aguilar, G., Zavaleta-Gatica, R. and Tiznado-Hernandez, M. (2007a). Improving postharvest quality of mango 'Haden' by UV-C treatment. *Postharvest Biology and Technology* 45:108–16.

Gonzalez-Aguilar, G. A., Villegas-Ochoa, M. A., Martinez-Téllez, M., Gardea, A. and Ayala-Zavala, J. F. (2007b). Improving antioxidant capacity of fresh-cut mangoes treated with UV-C. *Journal of Food Science* 72:S197–202.

Gouado, I., Schweigert, F. J., Ejoh, R. A., Tchouanguep, M. F. and Camp, J. V. (2007). Systemic levels of carotenoids from mangoes and papaya consumed in three forms (juice, fresh and dry slice). *European Journal of Clinical Nutrition* 61, 1180–8.

Grassmann, J., Hippeli, S. and Elstner, E. F. (2002). Plant's defence and its benefits for animals and medicine: role of phenolics and terpenoids in avoiding oxygen stress. *Plant Physiology and Biochemistry* 40:471–8.

Halliwell, B. (2007). Dietary polyphenols: Good, bad, or indifferent for your health? *Cardiovascular Research* 73:341–7. DOI:10.1016/j.cardiores.2006.10.004.

Halliwell, B. (2013). The antioxidant paradox: less paradoxical now? *British Journal of Clinical Pharmacology* 75:637–44.

Hulshof, P. J., Xu, C., van de Bovenkamp, P., Muhilal, A. and West, C. E. (1997). Application of a validated method for the determination of provitamin A carotenoids in Indonesian foods of different maturity and origin. *Journal of Agricultural and Food Chemistry* 45:1174–9.

Ibarra-Garza, I., Ramos-Parra, P. A., Hernandéz-Brenes, C. and Jacobo-Velázquez, D. A. (2015). Effects of postharvest ripening on the nutraceutical and physicochemical properties of mango (*Mangifera indica* L. cv Keitt). *Postharvest Biology and Technology* 103:45–55.

Jahurul, M., Zaidul, I., Ghafoor, K., Al-Juhaimi, F. Y., Nyam, K.-L., Norulaini, N., Sahena, F. and Omar, A. M. (2015). Mango (*Mangifera indica* L.) by-products and their valuable components: A review. *Food Chemistry* 183:173–80.

Jaleel, C. A., Riadh, K., Gopi, R., Manivannan, P., Ines, J., Al-Juburi, H. J., Chang-Xing, Z., Hong-Bo, S. and Panneerselvam, R. (2009). Antioxidant defense responses: physiological plasticity in higher plants under abiotic constraints. *Acta Physiologiae Plantarum* 31:427–36.

Jiang, Z., Zheng, Y., Qiu, R., Yang, Y., Xu, M., Ye, Y. and Xu, M. (2015). Short UV-B exposure stimulated enzymatic and nonenzymatic antioxidants and reduced oxidative stress of cold-stored mangoes. *Journal of Agricultural and Food Chemistry* 63:10965–72. DOI:10.1021/acs.jafc.5b04460.

Joas, J., Vulcain, E., Desvignes, C., Morales, E. and Léchaudel, M. (2012). Physiological age at harvest regulates the variability in postharvest ripening, sensory and nutritional characteristics of mango (*Mangifera indica* L.) cv. Coghshall due to growing conditions. *Journal of the Science of Food and Agriculture* 92:1282–90.

Junmatong, C., Faiye, B., Rotarayanont, S., Uthaibutra, J., Boonyakiat, D. and Saengnil, K. (2015). Cold storage in salicylic acid increases enzymatic and non-enzymatic antioxidants of Nam Dok Mai No. 4 mango fruit. *Scientia Asia* 41:12–21.

Kalyanaraman, B., Feix, J. B., Sieber, F., Thomas, J. P. and Girotti, A. W. (1987). Photodynamic action of merocyanine-450 on artificial and natural cell-membranes – involvment of singlet molecular oxygen. Proceedings of the National Academy of Sciences of the United States of America 84:2999–3003.

Kaushik, N., Kaur, B. P., Rao, S. and Mishra, H. N. (2014). Effect of high pressure processing on color, biochemical and microbiological characteristics of mango pulp (*Mangifera indica* cv. Amrapali). *Innovative Food Science and Emerging Technologies* 22:40–50.

Khaliq, G., Mohamed, M. T. M., Ghazali, H. M. Ding, P. and Ali, A. (2016). Influence of gum arabic coating enriched with calcium chloride on physiological, biochemical and quality responses of mango (*Mangifera indica* L.) fruit stored under low temperature stress. *Postharvest Biology and Technology* 111:362–9.

Kim, S. H., Moon, J. Y., Kim, H., Lee, D. S., Cho, M. and Choi, H. K. (2010). Antioxidant and antiproliferative activities of mango (*Mangifera indica* L.) flesh and peel. *Food Chemistry* 121:429–35.

Kim, S. H., Cho, S. K., Min, T. S., Kim, Y., Yang, S. O., Kim, H. S., Hyun, S. H., Kim, H. A., Kim, Y. S. and Choi, H. K. (2011). Ameliorating effects of mango (*Mangifera indica* L.) fruit on plasma ethanol level in a mouse model assessed with ^{1}H-NMR based metabolic profiling. *Journal of Clinical Biochemistry and Nutrition* 48:214–21.

Kim Y., Brecht J. K., Talcott S. T. (2007) Antioxidant phytochemical and fruit quality changes in mango (*Mangifera indica* L.) following hot water immersion and controlled atmosphere storage. *Food Chemistry* 105:1327–34.

Kim, Y., Lounds-Singleton, A. J. and Talcott, S. T. (2009). Antioxidant phytochemical and quality changes associated with hot water immersion treatment of mangoes (*Mangifera indica* L.). *Food Chemistry* 115:989–93.

Kondo, S., Kittikorn, M. and Kanlayanarat, S. (2005). Preharvest antioxidant activities of tropical fruit and the effect of low temperature storage on antioxidants and jasmonates. *Postharvest Biology and Technology* 36:309–18.

Krinsky, N. I. and Johnson, E. J. (2005). Carotenoid actions and their relation to health and disease. *Molecular Aspects of Medecine* 26:459–516.

Larrauri, J. A., Rupérez, P. and Saura-Calixto, F. (1997). Effect of drying temperature on the stability of polyphenols and antioxidant activity of red grape pomace peels. *Journal of Agricultural and Food Chemistry* 45:1390–3.

Léchaudel, M., Urban, L. and Joas, J. (2010). Chlorophyll fluorescence, a nondestructive method to assess maturity of mango fruits (Cv. 'Cogshall') without growth conditions bias. *Journal of Agricultural and Food Chemistry* 58:7532–8. DOI:10.1021/jf101216t.

Léchaudel, M., Lopez-Lauri, F., Vidal, V., Sallanon, H. and Joas, J. (2013). Response of the physiological parameters of mango fruit (transpiration, water relations and antioxidant system) to its light and temperature environment. *Journal of Plant Physiology* 170:567–76.

Lichtenthaler, H. K. and Buschmann, C. (2001). Chlorophylls and carotenoids: Measurement and characterization by UV-VIS spectroscopy. *Current Protocols in Food Analytical Chemistry*.

Linnewiel, K., Ernst, H., Caris-Veyrat, C., Ben-Dor, A., Kampf, A., Salman, H., Danilenko, M., Levy, J. and Sharoni, Y. (2009). Structure activity relationship of carotenoid derivatives in activation of the electrophile/antioxidant response element transcription system. *Free Radical Biology and Medicine* 47:659–67.

Lopes, M. M. A., Miranda, M. R. A., Moura, C. F. H. and Filho, J. E. (2012). Bioactive compounds and total antioxidant capacity of cashew apples (*Anacardium occidentale* L.) during the ripening of early dwarf cashew. *Ciência e Agrotecnologia* 36:325–32.

Lopes, M. M., Silva, E. O., Canuto, K. M., Silva, L. M., Gallao, M. I., Urban, L., Ayala-Zavala, J. F. and Miranda, M. R. A. (2015). Low fluence pulsed light enhanced phytochemical content and antioxidant potential of 'Tommy Atkins' mango peel and pulp. *Innovative Food Science & Emerging Technologies* 33, 216–24.

Luo, F., Ly, Q., Zhao, Y., Hu, G., Huang, G., Zhang, J., Sun, C., Li, X. and Chen, K. (2012). Quantification and purification of mangiferin from Chinese mango (*Mangifera indica* L.) cultivars and its protective effect on human umbilical vein endothelial cells under H_2O_2-induced Stress. *International Journal of Molecular Sciences* (13):11260–74. DOI:10.3390/ijms130911260.

Maciel, L. F., Oliveira, L. S., Bispo, E. S. and Miranda, M. P. S. (2011). Antioxidant activity, total phenolic compounds and flavonoids of mangoes coming from biodynamic, organic and conventional cultivations in three maturation stages. *British Food Journal* 13(9):1103–13.

Madsen, H. L. and Bertelsen, G. (1995). Spices as antioxidants. *Trends in Food Science & Technology* 6:271–7.

Manach, C., Williamson, G., Morand, C., Scalbert, A. and Rémésy, C. (2005). Bioavailability and bioefficacy of polyphenols in humans. I. Review of 97 bioavailability studies. *The American Journal of Clinical Nutrition* 81(1 Suppl):230S–42S. PMid:15640486.

Manthey, J. A. and Perkins-Veazie, P. (2009). Influences of harvest date and location on the levels of β-carotene, ascorbic acid, total phenols, the in vitro antioxidant capacity, and phenolic profiles of five commercial varieties of mango (*Mangifera indica* L.). *Journal of Agricultural and Food Chemistry* 57:10825–30.

Masibo, M. and He, Q. (2008). Major mango polyphenols and their potential significance to human health. *Comprehensive Reviews in Food Science and Food Safety* 7:309–19.

Mercadante, A. Z. and Rodriguez-Amaya, D. B. (1998). Effects of ripening, cultivar differences, and processing on the carotenoid composition of mango. *Journal of Agricultural and Food Chemistry* 46:128–30.

Mercadante, A. Z., Rodriguez-Amaya, D. B. and Britton, G. (1997). HPLC and mass spectrometric analysis of carotenoids from mango. *Journal of Agricultural and Food Chemistry* 45:120–3.

Mildel, J., Elstner, E. F. and Grabmann, J. (2007). Synergetic effects of phenolics and cpartenoids on human low-density lipoprotein oxidation. *Molecular Nutrition & Food Research* 51:956.

Molan, A. L., Liu, Z. and Plimmer, G. (2014). Evaluation of the effect of blackcurrant products on gut microbiota and on markers of risk for colon cancer in humans. *Phytotherapy Research* 28: 416–22.

Natal, D. I. G., Moreira, M. E. C., Militão, M. S., Benjamin, L. A., Dantas, M. I. S., Ribeiro, S. M. R. and Martino, H. S. D. (2016). Ubá mango juices intake decreases adiposity and inflammation in high-fat diet-induced obese Wistar rats. *Nutrition* (in press) 1–8.

Nisperos-Carriedo, M. O., Buslig, B. S. and Shaw, P. E. (1992). Simultaneous detection of dehydroascorbic, ascorbic, and some organic acids in fruits and vegetables by HPLC. *Journal of Agricultural and Food Chemistry* 40:1127–30.

Noctor, G., Veljovic-Jovanovic, S., Driscoll, S., Novitskaya, L. and Foyer, C. H. (2002). Drought and oxidative load in the leaves of C-3 plants: a predominant role for photorespiration? *Annals of Botany* 89:841–50.

Noratto, G. D., Bertoldi, M. C., Krenek, K., Talcott, A. T., Stringheta, P. C. and Mertens-Talcott, S. (2010). Anticarcinogenic effects of polyphenolics from mango (*Mangifera indica*) varieties. *Journal of agricultural and Food Chemistry* 58:4104–12. DOI:10.1021/jf903161g.

Normand, F., Lauri, P.-E. and Legave, J.-M. (2015). Climate change and its probable effects on mango production and cultivation, [ISHS Mango symposium. *Acta Horticulturae* 1075:21–31.

Obanda, M., Owuor, P. O. and Taylor, S. J. (1997). Flavanol composition and caffeine content of green leaf as quality potential indicators of Kenyan black teas. *Journal of the Science of Food and Agriculture* 74:209–15.

Oliveira, B. G., Costa, H. B., Ventura, J. A., Kondratyuk, T. P., Barroso, M. E. S., Correia, R. M., Pimentel, E. F., Pinto, F. E., Endringer, D. C. and Romão, W. Chemical profile of mango (*Mangifera indica* L.) using electrospray ionisation mass spectrometry (ESI-MS). *Food Chemistry* 149:253–63.

Oliveira, A. B. Moura, C. F. H., Filho, E. G., Marco, C. A., Urban, L. and Miranda, M. R. A. (2013). The impact of organic farming on quality of tomatoes is associated to increased oxidative stress during fruit development. *PLoS ONE* 8(2):e56354 DOI:doi:10.1371/journal.pone.0056354.

Olmedo, J. M., Yiannias, J. A., Windgassen, E. B. and Gornet, N. K. (2006). Scurvy: a disease almost forgotten. *International Journal of Dermatology* 45:909–13.

Ornelas-Paz, J. D. J., Yahia, E. M. and Gardea-Bejar, A. (2007). Identification and quantification of xanthophyll esters, carotenes, and tocopherols in the fruit of seven Mexican mango cultivars by liquid chromatography-atmospheric pressure chemical ionization-time-of-flight mass spectrometry [LC-(APcl+)-MS]. *Journal of Agricultural and Food Chemistry* 55:6628–35.

Osorio, S. and Fernie, A. R. (2013). Biochemistry of fruit ripening. In: Seymour, G. B., Poole, M., Giovannoni, J. J. and Tucker, G. A. (eds), *The Molecular Biology and Biochemistry of Fruit Ripening*. Blackwell Publishing Ltd., Iowa, pp. 1–19.

Palafox-Carlos, H., Ayala-Zavala, J. F. and González-Aguilar, G. A. (2011). The role of dietary fiber in the bioaccessibility and bioavailability of fruit and vegetable antioxidants. *Journal of Food Science* 76(1):R6–15. DOI:http://dx.doi.org/10.1111/j.1750-3841.2010.01957.x. PMid:21535705.

Palafox-Carlos, H., Yahia, E., Islas-Osuna, M. A., Gutierrez-Martinez, P., Robles-Sánchez, M. and González-Aguilar, G. A. (2012). Effect of ripeness stage of mango fruit (*Mangifera indica* L., cv. Ataulfo) on physiological parameters and antioxidant activity. *Scientia Horticulturae* 135:7–13.

Poiroux-Gonord, F., Fanciullino, A.-L., Poggi, I. and Urban, L. (2012a). Carbohydrate control over carotenoid build-up is conditional on fruit ontogeny in clementine fruits. *Physiologia Plantarum*:n/a-n/a. DOI:10.1111/j.1399–3054.2012.01672.x.

Poiroux-Gonord, F., Fanciullino, A.-L., Bert, L. and Urban, L. (2012b). Effect of fruit load on maturity and carotenoid content of clementine (*Citrus clementina* Hort. ex Tan.) fruits. *Journal of the Science of Food and Agriculture* 92:2076–83. DOI:10.1002/jsfa.5584.

Poiroux-Gonord, F., Bidel, L. P. R., Fanciullino, A.-L., Gautier, H., Lauri-Lopez, F. and Urban, L. (2010). Health benefits of vitamins and secondary metabolites of fruits and vegetables and prospects to increase their concentrations by agronomic approaches. *Journal of Agricultural and Food Chemistry* 58:12065–82. DOI:10.1021/jf1037745.

Poiroux-Gonord, F., Santini, J., Fanciullino, A.-L., Lopez-Lauri, F., Giannettini, J., Sallanon, H., Berti, L. and Urban, L. (2013). Metabolism in orange fruits is driven by photooxidative stress in the leaves. *Physiologia Plantarum* 149, 175–87. DOI:10.1111/ppl.12023.

Pott, I., Breithaupt, D. E. andCarle, R. (2003). Detection of unusual carotenoid esters in fresh mango (*Mangifera indica* L. cv. 'Kent'). *Phytochemistry* 64:825–9.

Prasanna, V., Prabha, T. N. and Tharanathan, R. N. (2007). Fruit ripening phenomena: an overview. *CRC Critical Reviews in Food Science* 47:1–19.

Raskin, I. and Ripoll, C. (2004). Can an apple a day keep the doctor away? *Current Pharmaceutical Design* 10:3419–29.

Razzaq, K., Khan, A. S., Malik, A. U. and Shahid, M. (2013). Ripening period influences fruit softening and antioxidative system of 'Samar Bahisht Chaunsa' mango. *Scientia Horticulturae* 160:108–14.

Reyes, L. F. and Cisneros-Zevallos, L. (2007). Electron-beam ionizing radiation stress effects on mango fruit (*Mangifera indica* L.) antioxidant constituents before and during postharvest storage. *Journal of Agricultural and Food Chemistry* 55:6132–9.

Ribeiro, S. M. R. and Schieber, A. (2010). *Bioactive foods in promoting health: fruit s and vegetables*. Bioactive Compounds in Mango (*Mangifera indica* L.), Academic Press, London.

Ribeiro, S. M. R., Queiroz, J. H., de Queiroz, M. E. L. R., Campos, F. M. and Sant'Ana, H. M. P. (2007). Antioxidant in mango (*Mangifera indica* L.) pulp. *Plant Foods for Human Nutrition* 62:13–17.

Ripoll, J., Urban, L., Staudt, M., Lopez-Lauri, F., Bidel, L. P. and Bertin, N. (2014). Water shortage and quality of fleshy fruits, making the most of the unavoidable. *Journal of Experimental Botany*:eru197.

Robles-Sánches, M., Astiazarán-García, H., Belloso, O. M., Gorinstein, S., Parrilla, E. A., Rosa, A. L. de., Yepiz-Plascencia, G. and González-Aguillar, G. A. (2011). Influence of whole and fresh-cut mango intake on plasma lipids and antioxidant. *Food Research International* 44:1386–91.

Robles-Sánchez, R. M. Islas, O., Astiazáran-Garcia, H., Vásquez-Ortiz, O., Martín-Belloso, O., Gorinstein, S. and González-Aguillar, G. A. (2009). Quality index, consumer acceptability, bioactive compounds, and antioxidant activity of fresh-cut 'Ataulfo' mangoes (*Mangifera Indica* L.) as affected by low-temperature storage. *Journal of Food Science* 74(3): S126–S134. DOI: doi:10.1111/j.1750-3841.2009.01104.x.

Safitri, A., Theppakorn, T., Naradisorn, M. and Setha, S. (2015). Effects of UV-C irradiation on ripening quality and antioxidant capacity of mango fruit cv. Nam Dok Mai Si Thong. *Journal of Food Science and Agricultural Technology* (JFAT) 1:164–70.

Saura-Calixto, F., Serrano, J. and Goñi, I. (2007). Intake and bioaccessibility of total polyphenols in a whole diet. *Food Chemistry* 101(2):492–501. DOI:http://dx.doi.org/10.1016/j.foodchem.2006.02.006.

Schaffer, B., Urban, L., Lu, P. and Whiley, A. (2009). 6 Ecophysiology. *The Mango: Botany, Production and Uses*, p. 170.

Schieber, A., Ullrich, W. and Carle, R. (2000). Characterization of polyphenols in mango puree concentrate by HPLC with diode array and mass spectrometric detection. *Innovative Food Science & Emerging Technologies* 1:161–6.

Septembre-Malaterre, A., Stanilas, G., Douraguia, E. and Marie-Paule, G. (2016). Evaluation of nutritional and antioxidant properties of the tropical fruits banana, litchi, mango, papaya, passion fruit and pineapple cultivated in Réunion French Island. *Food Chemistry* 212:225–33.

Setiawan, B., Sulaeman, A., Giraud, D. W. and Driskell, J. A. (2001). Carotenoid content of selected Indonesian fruits. *Journal of Food Composition and Analysis* 14:169–76.

Shi, S., Ma, X., Xu, Y., Wu, H. and Wang, S. (2015). Evaluation of 28 mango genotypes for physicochemical characters, antioxidant capacity, and mineral content. *Journal of Applied Botany and Food Quality* 88:264–73. DOI:10.5073/JABFQ.2015.088.039.

Shivashankara, K., Isobe, S., Al-Haq, M. I., Takenaka, M. and Shiina, T. (2004). Fruit antioxidant activity, ascorbic acid, total phenol, quercetin, and carotene of Irwin mango fruits stored at low temperature after high electric field pretreatment. *Journal of Agricultural and Food Chemistry* 52:1281–6.

Sies, H. and Menck, C. F. M. (1992). Singlet oxygen induced DNA damage. *Mutation Research* 275:367–75.

Smirnoff, N. (1993). The role of active oxygen in the response of plants to water-deficit and desiccation. *New Phytologist* 125:27–58.

Sogi, D., Siddiq, M., Roidoung, S. and Dolan, K. (2012). Total phenolics, carotenoids, ascorbic ccid, and antioxidant properties of fresh-cut mango (*Mangifera indica* L., cv. Tommy Atkin) as affected by infrared heat treatment. *Journal of Food Science* 77:C1197–202.

Sreelakshmi, Y. and Sharma, R. (2008). Differential regulation of phenylalanine ammonia lyase activity and protein level by light in tomato seedlings. *Plant Physiology and Biochemistry* 46:444–51.

Surh, Y.-J. and Na, H.-K. (2008). NF-kB and Nrf2 as prime molecular targets for chemoprevention and cytoprotection with anti-inflammatory and antioxidant phytochemicals. *Genes & Nutrition* 2:313–17.

Talcott, S. T., Moore, J. P., Lounds-Singleton, A. J. and Percival, S. S. (2005). Ripening associated phytochemical changes in mangos (*mangifera indica*) following thermal quarantine and low-temperature storage. *Journal of Food Science* 70:C337–41.

Tian, S., Liu, J., Zhang, C. and Meng, X. (2010). Quality properties of harvested mango fruits and regulating technologies. New trends in postharvest management of fresh produce II. *Fresh Produce* 4:49–54.

Urban, L. and Léchaudel, M. (2005). Effect of leaf-to-fruit ratio on leaf nitrogen content and net photosynthesis in girdled branches of *Mangifera indica* L. *Trees-Structure and Function* 19:564–71. DOI:10.1007/s00468-005-0415-6.

Urban, L. and Alphonsout, L. (2007). Girdling decreases photosynthetic electron fluxes and induces sustained photoprotection in mango leaves. *Tree Physiology* 27:345–52.

Urban, L., Charles, F., de Miranda, M. R. A. and Aarrouf, J. (2016). Understanding the physiological effects of UV-C light and exploiting its agronomic potential before and after harvest. *Plant Physiology and Biochemistry.*

Valderrain-Rodríguez, G., Quirós-Sauceda, A. Mercado-Mercado, G., Ayala-Zavala, J., Astiazarán-García, H., Robles-Sánchez, R. M., Wall-Medrano, A., Sayago-Ayerdi S., and González-Aguillar, G. A. (2016). Effect of dietary fiber on the bioaccessibility of phenolic compounds of mango, papaya and pineapple fruits by an *in vitro* digestion model. *Food Science and Technology.* DOI:http://dx.doi.org/10.1590/1678-457X.6729.

Veda, S., Platel, K. and Srinivasan, K. (2007). Varietal differences in the bioaccessibility of β-carotene from mango (*Mangifera indica*) and papaya (*Carica papaya*) fruits. *Journal of Agricultural and Food Chemistry* 55:7931–5.

Vinci, G., Botrè, F., Mele, G. and Ruggieri, G. (1995). Ascorbic acid in exotic fruits: a liquid chromatographic investigation. *Food Chemistry* 53:211–14.

Wang, L. S. Burke, C. A., Hasson, H., Kuo, C. T., Molmenti, C. L. S., Seguin, C., Liu, P., Huang, T. H. M., Frankel, W. L. and Stoner, G. D. (2014). A phase Ib study of the effects of black raspberries on rectal polyps in patients with familial adenomatous polyposis. *Cancer Prevention Research* 7:666674.

Wenkam, N. (1979). Nutritional aspects of some tropical plant foods. *Tropical Foods: Chemistry and Nutrition* 2:341–50.

Zhao, Z., Jiang, W., Cao, J., Zhao, Y. and Gu, Y. (2006). Effect of cold-shock treatment on chilling injury in mango (*Mangifera indica* L. cv. Wacheng') fruit. *Journal of the Science of Food and Agriculture* 86:2458–62.

Chapter 3

Genetic engineering of tomato to improve nutritional quality, resistance to abiotic and biotic stresses, and for non-food applications

B. Kaur and A. K. Handa, Purdue University, USA; and A. K. Mattoo, USDA-ARS, USA

1 Introduction

Genetic engineering is among the fastest adopted crop technologies in modern history with global hectarage increasing from 1.7 million in 1996 to 179.7 million in 2015 (James, 2015, ISAAA). Successful transformation of plant cells started with genes of bacterial origin (in 1983) involving academicians in the United States and Europe as well as biotechnology companies, including Monsanto (Bevan et al., 1983; Fraley et al., 1983; Hererra-Estrella et al., 1983; Murai et al., 1983). The first genetically engineered food with superior shelf life trait released in the market was the Flavr Savr tomato in 1994 (Kramer and Redenbaugh, 1994; Kramer et al., 1990, 1992). The standardized transformation protocols and ease of tissue culture for tomato led to genetic engineering for specialized traits and process development of a great number of new lines evidenced by 786 instances of notifications and permits for field trials between 1986 and 2016 (Virginia Tech University, 2016, Information Systems for Biotechnology).

http://dx.doi.org/10.19103/AS.2016.0007.10

Tomato (*Solanum lycopersicum* L.) has thus become an excellent research model for elucidation of fundamental physiological processes, molecular genetics, development and pathology in general, and fruit development and ripening studies in climacteric fruits in particular (Meissner et al., 1997; Fatima et al., 2008; Klee and Giovannoni, 2011; Upadhyay et al., 2013, 2014; Anwar et al., 2015). The deciphering of the genome of inbred tomato cultivar, Heinz 1706 (Sato et al., 2012, The Tomato Genome Consortium) and multitude of genetic resources together with established regulatory framework well in place have made tomato a model fruit for genetic dissection. In addition, fruit ripening mutants (Giovannoni, 2004, 2007) and genetic linkages for fruit quality have culminated into a very clear road map for scientists to further unravel the intricacies of genes governing fruit quality attributes as well as fundamental metabolic processes. The relatively small (950 Mb) tomato genome is organized into 12 chromosomes comprising 34727 genes that encode proteins, of which 30855 genes are supported/validated by RNA sequencing data (Sato et al., 2012).

A member of the Solanaceae family which contains more than 3000 species, including equally and economically important crops such as potato, eggplant, tobacco, petunia and pepper (Bai and Lindhout, 2007), tomato is the second most consumed vegetable next to potato (FAOSTAT, 2015). Mainland China led the average yearly global production of tomatoes from 1961 to 2014 at 14.6 Million Tonnes (MT) followed by the United States at an average yearly production of 9.2 MT (FAOSTAT, 2015). Fresh and processed tomatoes account for returns of more than $2 billion annually in the United States (Economic Research Service, USDA, 2016). In light of the potential anti-cancer and anti-oxidative properties of lycopene, ß-carotene and flavonoids, and phytonutrients abundant in tomato, the production and consumption of tomato are projected to rise every year (Raiola et al., 2014). A high global demand for tomatoes, particularly in the processing industry, requires improvements in economically important agronomic traits (Fatima et al., 2013).

High-efficiency transformation and a reproducible regeneration protocol are central to functional genomics studies with this important vegetable crop. Synthetic biology has given impetus to engineering genes for introduction of new traits and biochemical pathways into crop plants of interest (Lu et al., 2013). In addition, co-introduction or pyramiding of multiple genes into vector constructs for transformation has been demonstrated to be efficacious for engineering metabolic pathways and resistance against pathogens in plants (Ye et al., 2000; Zhao et al., 2003; Abdeen et al., 2005).

This chapter provides comprehensive information on genetic engineering studies that have introduced beneficial traits in tomato. The vast bibliographic database was made possible to view through Scopus and Purdue University library (www.lib.purdue. edu), to sift through literature from 1983 to 2016. Using keywords such as tomato, genetic engineering, transformation, *Agrobacterium*-mediated, fruit quality, biotic stress tolerance, abiotic stress tolerance, salinity and many others resulted in the listing of hundreds of relevant publications. However, not all documents listed dealt with genetic engineering studies. Many reviews on genetic engineering and tomato serve as valuable resource in refining this chapter (Razdan and Mattoo, 2006; Fatima et al., 2008; Handa et al., 2010, 2012; Pandey et al., 2011; Khaliluev and Shpakovskii, 2013; Bergougnoux, 2014; Nath et al., 2014). Herein we focus on tomato transformed for introduction and manipulation of transgenes for four broad categories: improved fruit quality and enhancement of shelf life, abiotic stress tolerance, biotic stress tolerance and production of oral vaccines.

2 History of tomato transformation and challenges

Attempts to modify tomato started with egg transformation using irradiated pollen but failed (Sanford et al., 1984). Soon thereafter, the first success that heralded genetic transformation of tomato involved *Agrobacterium tumefaciens* as the carrier and tomato leaf disks (McCormick et al., 1986). Various explants such as cotyledons (Abu-El-Heba et al., 2008; Kaur and Bansal, 2010), epicotyls, hypocotyls (Moghaieb et al., 2004), stems (Ma et al., 2015), petioles (Sigareva et al., 2004), internodes (Chyi and Phillips, 1987) and leaves (Agharbaoui et al., 1995) have been used ever since to develop transformation and regeneration methods for different tomato cultivars. Factors that need to be considered while embarking on tomato transformation include age of the explant (Davis et al., 1991), type of the explant (Sigareva et al., 2004; Hasan et al., 2008), size of the explant (Chaudry et al., 2010; Ajenifujah-Solebo et al., 2012) and the cultivar to be used (Ellul et al., 2003; Cortina and Culianez-Macia, 2004; Ume-e-Ammara et al., 2014). In addition, other factors such as using nurse cells or feeder layers, complex media combinations (Wu et al., 2011a), subculture frequency, adding plant growth regulators (Ume-e-Ammara et al., 2014), inclusion of acetosyringone, type and concentration of antibiotics as marker, co-cultivation time and plasmid vector construction (Yasmeen et al., 2009) are variables that can impact the establishment of a robust and reliable transformation regimen for tomato. Higher shoot organogenesis was obtained from hypocotyl explants rather than cotyledons from three tomato varieties and the addition of thidiazuron enhanced their shoot differentiation (Murlidhar Rao et al., 2007).

Efficient transformation methods have been reported for defensin gene (El-Siddig et al., 2011) and coat protein of tomato yellow leaf curl virus using different media formulations that particularly included growth regulators such as putrescine, zeatin riboside and indole acetic acid (Ume-e-Ammara et al., 2014). An efficient *Agrobacterium*-mediated transformation protocol on tomato cotyledons enabled studies on the effect of different variables such as seed germination medium, seedling age, pre-culture duration, co-cultivation medium, pH of medium, kanamycin concentration and tobacco feeder cell layer on transforming three tomato cultivars (Rai et al., 2012). Thus, pre-culturing the explant for 6 days after a 5-min inoculation with *Agrobacterium* culture in MS medium fortified with 8.9 µM 6-benzyladenine, 9.3 µM kinetin and 0.4 mgL⁻¹thiamine, pH 5.0, was beneficial for obtaining high transformation frequency with a number of tomato cultivars (Rai et al., 2012). Notably, the layering of tobacco feeder cell contributed little to the transformation efficiency of the few tomato cultivars studied.

Effects of stage of explants (stem from 4–5-day-old seedling and cotyledon from 8–9-day-old seedling), pre-culture duration (3 days for stem and 2 days for cotyledon), *Agrobacterium* density (OD_{600} = 0.6 for both explants), infection time (15 min for stem and 20 min for cotyledon) and co-cultivation duration (4 days for stem and 3 days for cotyledon) were analysed to develop a high-throughput transformation protocol for transformation of *Crocus sativus* zeaxanthin 7,8-cleavage dioxygenase gene in two Chinese tomato cultivars, Zheza No.905 and Shengya (Ma et al., 2015). Transformation efficiency of cotyledon and stem explants was 26.33% and 28.00% for Zheza No.905 and 19.33% and 23.33% for Shengya, respectively.

3 Genetic engineering of tomato for fruit quality and shelf life

Fruits, derived from different parts of a flower, are highly diverse in their structure and physiological functions (Handa et al., 2012). Fruit quality attributes such as long post-harvest shelf life, attractive colour, large size, high nutritive value, improved palatability and optimum rheological properties for processing are economically advantageous for farmers and supply chain personnel alike with a value for the money of the consumers.

The maiden genetically engineered tomato, Flavr Savr (CGN-89564), with reduced polygalacturonase (PG) expression via antisense RNA technology (later found to be mediated by interfering RNAs, Krieger et al., 2008), enabled superior juice viscosity and shelf life, and was commercialized over 20 years ago (Kramer and Redenbaugh, 1994; Kramer et al., 1990, 1992). It paved the way for experimentalists and researchers to tap into numerous genes governing tomato fruit quality attributes for use in the genetic manipulation to develop better quality tomatoes in terms of shape, size, texture, phytonutrient levels and volatiles (selected examples in Table 1).

Fruit architecture, size and shape in tomato are inherited by cumulative gene action. Fruit shape variation has been attributed to mutations found in one of the following four genes, *SUN*, *OVATE*, *LOCULE NUMBER* (*lc*) and *FASCIATED* (*fas*) (Rodriguez et al., 2011). The former two genes regulate fruit elongation while the latter ones govern locule number and flatness of fruit. Overexpression of *SUN* driven by CaMV *35S* (Cauliflower mosaic virus 35S) promoter led to the development of a parthenocarpic elongated fruit due to changes in cell division pattern (Wu et al., 2011b). These fruit had increased cell number longitudinally and decreased in those in the transverse direction in the developing fruit. *SUN* expression was hypothesized to perturb auxin levels but this remains to be demonstrated. Overexpression of *IQD12* (a protein containing an IQ67 domain consisting of multiple IQxxxRGxxxR motifs) at the *SUN* locus increased fruit elongation, while silencing it by RNAi decreased fruit elongation (Xiao et al., 2008). *OVATE* is a repressor of transcription and leads to reduced fruit size; however, it determines fruit shape pattern prior to anthesis (Bohner and Bangerth, 1988; Liu et al., 2002; Monforte et al., 2014). Overexpression of *OFP1* or ectopic overexpression of *OVATE* under the control of CaMV *35S* promoter reduced fruit elongation in tomato (Ku et al., 1999) or produced round fruit (Liu et al., 2002). *FAS*, expressed early during the development of stamens and carpels, is a transcription factor belonging to the YABBY family of genes (involved in leaf, flower and fruit development), nine of which have been characterized in tomato both *in silico* and experimentally (Han et al., 2015). *FAS* and *lc* exhibit an epistatic interaction with *fas* having a stronger effect by increasing the locules' number from 2 to 6 while *lc* increases the number to 3 or 4 (Lippman and Tanksley, 2001). Two SNPs located in *lc* locus were attributed with function to increase locule number (Muños et al., 2011). Map-based cloning placed the two SNPs downstream of WUS (WUSCHEL, homeobox transcription factor) gene (required for shoot and floral integrity). The final fruit size is determined by cell expansion via endoreduplication, a form of nuclear polyploidization (Cheniclet et al., 2005), which is induced by cyclin-dependent kinase inhibitors such as *WEE1* (Sun et al., 1999). Ploidy levels along with fruit mass, growth

Table 1 Genetic engineering of tomato for improved/altered fruit quality

Gene and source	Description	Trait/phenotype conferred	References
D-galacturonate reductase *FaGalUR* *Fragaria* x *ananassa* (strawberry)	Ectopic expression of *FaGalUR* gene driven by constitutive CaMV *35S* (Cauliflower mosaic virus *35S* promoter) and fruit-specific polygalacturonase promoter	Increased vitamin C content and antioxidant capacity	Amaya et al., 2015
Homogentisate phytyltransferase (*HPT*), tocopherol cyclase (*TCY*) and γ-tocopherol methyltransferase (TMT) *Synechocystis* sp. (cyanobacteria)	Plastid genome engineering by assembly of monocistronic expression cassettes for HPT, TCY and TMT in plastid expression vector pHK20	Expression of tocochromanol biosynthesis in chloroplasts and chromoplasts for high vitamin E activity in tomato	Lu et al., 2013
Lycopene β-cyclase (*Lycb-1*) (*Citrus*)	Overexpression of *Lcyb-1* driven by CaMV *35S*	Fourfold increase in β-carotene and increased total carotenoids	Guo et al., 2012
9-*cis*-Epoxycarotenoid dioxygenase 3 (*NCED1*) *Solanum lycopersicum*	Fruit-specific silencing (E8) of *SlNCED1* by RNAi	Increased accumulation of β-carotene and lycopene, decrease in abscisic acid	Sun et al., 2012
Spermidine synthase (*MdSPDS*1) *Malus* x *domestica* (apple)	Overexpression of *MdSPDS1* driven by CaMV *35S*	Increased lycopene content	Neily et al., 2011
TOMATO AGAMOUS-LIKE 1 *TAGL1* (*S. lycopersicum*)	Overexpression of (*TAGL1-SRDX*) under fruit ripening-specific tomato E8 promoter	Decrease in lycopene and isoprenoids	Itkin et al., 2009
TAGL1 *S. lycopersicum*	Overexpression of *TAGL1* driven by CaMV *35S*	Increase in lycopene and naringenin chalcone	Itkin et al., 2009
UV-DAMAGED DNA-BINDING PROTEIN 1 *DDB1* *S. lycopersicum*	Fruit-specific RNAi-mediated repression of *DDB1* driven by *E8* promoter	Increased pigment accumulation by virtue of increased plastid compartment space	Wang et al., 2008

(Continued)

Table 1 (*Continued*)

Gene and source	Description	Trait/phenotype conferred	References
Geraniol synthase *GES* *Ocimum basilicum*	Overexpression of *GES* under fruit ripening-specific tomato polygalacturonase promoter	Increase in carotenoid-derived aroma volatiles but decrease in carotenoids like phytoene, lycopene and β-carotene	Davidovich-Rikanati et al., 2007
Phytoene synthase *SlPsy-1* *S. lycopersicum*	Overexpression of *Psy-1* driven by CaMV *35S*	1.2-fold increase in total carotenoids, 1.3-fold increase in β-carotene, 2.3-fold increase in phytoene, 1.8 fold increase in phytofluene, decrease in phenylpropanoids and flavonoids	Fraser et al., 2007
Lycopene β-cyclase *crtY Erwinia herbicola* or *carRA Phycomyces blakesleeanus*	Overexpression of *crtY* or *carRA* driven by atpI (ATPase IV subunit) tobacco plastid-specific promoter	Fourfold increase in β-carotene accumulation, slight decrease in lycopene and total carotenoids	Wurbs et al., 2007
Fibrillin *FIB1, FIB2* *Capsicum annuum*	Overexpression of capsicum fibrillin genes driven by their own promoter	Twofold increase in carotenoids, for example 118% increase in lycopene, 64% increase in β-carotene, 36% increase in β-ionone, 74% increase in β-cyclocitral, 50% increase in citral, 122% increase in 6-methyl-5-hepten-2-one and 223% increase in geranylacetone	Simkin et al., 2007
1-Deoxy-D-xyluose-5-phosphate synthase *DXS* *Escherichia coli*	Expression of *DXS* driven by CaMV *35S* or fibrillin promoter	1.6-fold increase in total carotenoids, 2.4-fold and 2.2-fold increase in phytoene and β-carotene, respectively	Enfissi et al., 2005
Cryptochrome 2 *CRY2* *S. lycopersicum*	Overexpression of *CRY2* driven by CaMV *35S*	1.5-fold increase in lutein, 1.7-fold increase in total carotenoids and 2.9-fold increase in flavonoids accumulation	Giliberto et al., 2005

Genetic engineering of tomato   65

Table 1 (*Continued*)

Gene and source	Description	Trait/phenotype conferred	References
DE-ETIOLATED 1 *DET1* S. *lycopersicum*	Fruit-specific RNAi-mediated inhibition of *DET1* driven by P119, 2A11 and TFM7 promoters	Twofold increase in lycopene, fourfold increase in β-carotene and 3.5-fold increase in flavonoids	Davuluri et al., 2005
Lycopene β-cyclase *Lyc-b* S. *lycopersicum*	Overexpression of *Lyc-b* driven by CaMV *35S*	31.7-fold increase in β-carotene	D'Ambrosio et al., 2004
ELONGATED HYPOCOTYL 5 *HY5* S. *lycopersicum*	RNAi-mediated repression of *HY5* driven by CaMV *35S*	Decrease in total carotenoids	Liu et al., 2004
CONSTITUTIVELY PHOTOMORPHOGENIC 1 *COP1*-like S. *lycopersicum*	RNAi-mediated repression of *COP1*-like driven by CaMV *35S*	Twofold increase in total carotenoids	Liu et al., 2004
Lycopene β-cyclase β-*Lcy Arabidopsis thaliana* carotene β-hydroxylase *b-Chy C. annuum*	Overexpression of β-*Lcy* and *b-Chy* genes driven by tomato phytoene desaturase promoter	12-fold increase in β-carotene and 10-fold increase in total xanthophyll	Dharmapuri et al., 2002
Phytoene synthase *CrtB Erwinia uredovora*	Fruit-specific expression driven by polygalacturonase promoter	2.4-fold increase in phytoene, 1.8-fold increase in lycopene and 2.2-fold increase in β-carotene	Fraser et al., 2002
SAM decarboxylase *SPE2 Saccharomyces cerevisiae*	Overexpression of *SPE2* driven by fruit-specific E8 promoter	Two–threefold increase in lycopene, increase in transcripts related to flavonoid biosynthesis genes	Mehta et al., 2002; Mattoo et al., 2007
Phytoene desaturase *Crtl E. uredovora*	Overexpression of *Crtl* driven by CaMV *35S*	Threefold increase in β-carotene but decrease in lycopene and phytoene	Römer et al., 2000
Lycopene β-cyclase *SpB S. pennellii*	Overexpression of *SpB* driven by CaMV *35S*	Greater than sixfold increase in β-carotene but 1.8-fold decrease in lycopene	Ronen et al., 2000

(*Continued*)

Table 1 (*Continued*)

Gene and source	Description	Trait/phenotype conferred	References
SpB *S. pennellii*	Antisense-mediated downregulation of *SpB* driven by CaMV *35S*	Greater than sixfold decrease in β-carotene and a slight increase in lycopene	Ronen et al., 2000
β-Lcy *A. thaliana*	Overexpression of *β-Lcy* gene driven by tomato phytoene desaturase promoter	Greater than sixfold increase in β-carotene	Rosati et al., 2000
β-Lcy *S. lycopersicum*	Antisense-mediated downregulation of *β-Lcy* driven by tomato phytoene desaturase promoter	1.3-fold increase in lycopene; 1.7-fold increase in lutein; 50% decrease in *β-Lcy* expression	Rosati et al., 2000
SlPsy-1 *S. lycopersicum*	Overexpression of *Psy-1* driven by CaMV *35S*	High lycopene content accompanied by decrease in plant height and 30-fold decrease in gibberellin GA_1	Fray et al., 1995
Pectin methylesterase *PME* *S. lycopersicum*	Antisense-mediated downregulation of PME driven by CaMV *35S*	Increased juice and serum viscosity, higher precipitate weight ratio, increased size and degree of pectin methoxylation	Tieman et al., 1992; Thakur et al., 1996a, 1996b
Lipoxygenase B *LoxB* *S. lycopersicum*	Cosuppression under CaMV *35S*	Impaired MeJA production, altered metabolome and aminome	Kausch et al., 2012
viscosity 1 *vis1* *S. lycopersicum*	Silencing of *vis1* gene by RNAi technology	Enhanced ripening qualities under heat stress	Metwali et al., 2015
Spermidine synthase *SPE3* *S. cerevisiae*	Overexpression of *SPE3* driven by CaMV *35S*	Fruit shelf life, shrivelling and delayed decay, increased lycopene content	Nambeesan et al., 2010
Anthocyanin1 *ANT1* *S. chilense*	Overexpression of *ANT1* driven by CaMV *35S*	Increased anthocyanadins (petunidin, malvidin, delphinidin) in fruit	Schreiber et al., 2012

Table 1 (*Continued*)

Gene and source	Description	Trait/phenotype conferred	References
Delila *Del* *Antirrhinum majus*	Overexpression of *Del* driven by CaMV *35S*	23-fold increase in anthocyanins in mature leaves along with 40-fold and 50-fold increase in corolla and stamen, respectively, no change in fruit	Mooney et al., 1995
RP *Myc-rp* *Perilla fructescens*	Overexpression of *RP* driven by CaMV *35S*	Increase in anthocyanins in vegetative tissues and flowers	Gong et al., 1999
LC, a member of maize *R* gene family of MYC-type transcription factors *Zea mays*	Overexpression of *LC* driven by CaMV *35S*	Increase in anthocyanins in all vegetative tissues	Goldsbrough et al., 1996
Chalcone isomerase *Chi-A* *Petunia hybrida*	Overexpression of *Chi-A* driven by CaMV *double35S*	78-fold increase in peel flavonols accumulation primarily rutin	Muir et al., 2001
Transcription factors *C1* (MYB-type) and *LC* (MYC-type) *Z. mays*	Overexpression of *C1* and *LC* driven by fruit-specific E8 or CaMV *double35S*	Induced flavonoid synthesis in fruit flesh with a 10-fold increase in total flavonoids and 20-fold increase in total flavonols, primarily kaempferol	Bovy et al., 2002
ANT1 *S. lycopersicum*	Overexpression of *SlANT1* driven by CVM (Cassava vein mosaic promoter)	500-fold increase in anthocyanin accumulation	Mathews et al., 2003
Chalcone synthase *Chs1* *S. lycopersicum*	RNAi-mediated repression of *Chs1* driven by CaMV *d35S*	Decrease in total flavonoids, parthenocarpic fruit	Schijlen et al., 2007
Stilbene synthase *StSy* *Vitis vinifera* (grapes)	Overexpression of *StSy* driven by CaMV *35S*	Increased accumulation stilbenes (resveratrol and piceid) and naringenin chalcone, rutin	Schijlen et al., 2006
Chalcone synthase (*Chs1*) *P. hybrida* Chalcone reductase (*CHR*) *Medicago sativa*	Overexpression of *Chs1* and *CHR* driven by CaMV *35S*	Increase in butein and isoliquiritigenin accumulation along with naringenin chalcone and rutin	Schijlen et al., 2006

(*Continued*)

Table 1 (*Continued*)

Gene and source	Description	Trait/phenotype conferred	References
Chalcone isomerase (*CHI*) *P. hybrida* Flavone synthase (*CYP93B2*) *Gerbera hybrida*	Overexpression of *CHI* and *CYP93B2* driven by CaMV *35S*	16-fold increase in rutinflavonol, increased accumulation of luteolin-7-glucoside, luteolinaglycon, quercetin glycosides, naringenin chalcone and rutin	Schijlen et al., 2006
Stilbene synthase (*StSy*) *V. vinifera*	Overexpression of *StSy* driven by CaMV *35S*	Increase in *trans*-resveratrol (48.48 mg kg^{-1} fresh weight), *trans*-piceid (126.58 mg kg^{-1} fresh weight), twofold decrease in rutin, 2.4-fold decrease in naringenin, seedless fruit	Giovinazzo et al., 2005; Nicoletti et al.,2007
Cullin 4 (*CUL4*) *S. lycopersicum*	RNAi-mediated repression of *CUL4* driven by CaMV *35S*	Increase in anthocyanins accumulation and carotenoids (twofold increase in lycopene)	Wang et al., 2008
Isoflavone synthase (*IFS2*) *Glycine max*	Overexpression of *IFS2* driven by CaMV *35S*	Increased accumulation of genistin in leaves and naringenin chalcone in fruit peel	Shih et al., 2008
Rosea1 and *Delila* *A. majus*	Overexpression of *AmRos1* and *Del* under fruit ripening-specific tomato E8 promoter	Increase in pericarp anthocyanins comparable to blueberries and blackberries	Butelli et al., 2008
MYB12 *S. lycopersicum*	Silencing of *SlMYB12* driven by CaMV *35S* by RNAi technology	Decrease in flavonoid pigment naringenin chalcone, *y*-like phenotype	Adato et al., 2009
MYB12 *S. lycopersicum*	Overexpression of *SlMYB12* driven by CaMV *35S*	Rescued colourless peel, y tomato mutant phenotype	Adato et al., 2009
Stilbene synthase *StSy* *V. vinifera*	Overexpression of grape StSy under control of fruit-specific promoter *TomLoxB*	Increased accumulation of resveratrol, *trans*-resveratrol and piceid	D'Introno et al., 2009
ω-3 fatty acid desaturase *FAD3Brassica napus* or/ and *FAD7* *S. tuberosum*	Overexpression of *FAD3* and *FAD7* driven by CaMV *35S*	Increase in C18 polyunsaturated fatty acids	Domínguez et al., 2010

Table 1 (*Continued*)

Gene and source	Description	Trait/phenotype conferred	References
3-Hydroxy-3-methyl-glutaryl CoA reductase *HMGR-1* A. thaliana	Overexpression of *HMGR-1* driven by CaMV *35S*	2.4-fold increase in total phytosterol	Enfissi et al., 2005
Salicylic acid methyltransferase *SAMT* S. lycopersicum	Overexpression of *SAMT* driven by FMV *35S* (Figwort mosaic virus)	123-fold increase in methyl salicylate	Tieman et al., 2010
MYB12 A. thaliana	Overexpression of At*MYB12* driven by CaMV *35S*	27-fold increase in chlorogenic acid, 26-fold increase in dicaffeoylquinic acid, 42-fold increase in tricaffeoylquinic acid, 67-fold increase in quercetinrutinoside, 593-fold increase in kaempferolrutinoside	Luo et al., 2008
Amino acid aromatic decarboxylase (*AADC1A*) S. lycopersicum	Overexpression of *AADC1A* driven by FMV *35S*	10-fold increase in 1-nitro-2-phenylethane, 2-phenylethanol and 2-phenylacetaldehyde	Tieman et al., 2006
Odorant 1 *ODO1* P. hybrida	Overexpression of *ODO1* driven by fruit-specific E8 promoter	No increase in phenylalanine-derived volatile compounds	Dal Cin et al., 2011
S-linalool synthase *LIS* Clarkia breweri	Overexpression of *LIS* driven by fruit-specific E8 promoter	Increase in S-linalool and 8-hydroxylinalool	Lewinsohn et al., 2001
α-Zingiberene synthase *ZIS* O. basilicum	Overexpression of *ZIS* under fruit ripening-specific tomato polygalacturonase promoter	Accumulation of high levels of α-zingiberene (224-1000 ng g^{-1} fresh weight) and other sesquiterpenes	Davidovich-Rikanati et al., 2008
Carotenoid cleavage dioxygenase *CCDIBS. lycopersicum*	Antisense-mediated downregulation of *CCDIB* driven by FMV *35S*	50% decrease in β-ionone, greater than 60% decrease in geranylacetone but no morphological alterations or changes in carotenoids	Simkin et al., 2004
Lipoxygenase Tom*LoxC* S. lycopersicum	Antisense-mediated downregulation of Tom*LoxC* driven by CaMV *35S*	1.5% decrease in hexanal, hexenal and hexanol	Chen et al., 2004

and seed size were reduced in tomato upon silencing WEE1 under the control of CaMV 35S by antisense technology (Gonzalez et al., 2007). A cell cycle switch, CCS52A, arrests cell division and promotes endoreduplication (Cebolla et al., 1999). Overexpression of CCS52A activated anaphase-promoting complex E3 ubiquitin ligase and led to increased tomato fruit size (Mathieu-Rivet et al., 2010).

Taste, flavour and texture are important fruit quality attributes and important candidates for modification. Taste of tomato fruit may vary from being sweet to acidic as a result of a delicate balance between sugars and organic acids. Thaumatin, a sweet-tasting protein, is produced by an African plant, Thaumatococcus daniellii Benth (Van der Wel and Loeve, 1972). Engineering of thaumatin gene in tomato produced fruits with an enhanced sweet taste and particularly a sweet-based aftertaste (Bartoszewski et al., 2003). Overexpression of α-zingiberene synthase and geraniol synthase genes from Ocimum basilicum driven by ripening-specific PG promoter resulted in tomatoes accumulating geraniol, an acyclic monoterpene (Davidovich-Rikanati et al., 2007, 2008). These fruits had a floral aroma but were deficient in nutrient content, including reduced carotenoid levels. Fruit-specific expression of yeast S-adenosylmethionine decarboxylase (ySAMdc) enabled ripening-specific accumulation of polyamines, spermidine and spermine; longer vine life; and higher levels of lycopene in tomato fruit (Mehta et al., 2002). These fruit had enhanced accumulation of glutamine, asparagine and organic acids in the red fruit with significant decreases in the contents of valine, aspartate, sucrose and glucose. Notably, these fruit were found to have an increased acid to sugar ratio, an attribute desired by geneticists for providing enhanced flavour and aftertaste to tomato (Mattoo et al., 2006, 2007).

Ethylene is a gaseous plant hormone pivotal to climacteric fruit ripening and softening (Oeller et al., 1991; Mattoo and Suttle, 1991; Fluhr and Mattoo, 1996). Biochemical pathways involving its biosynthesis, degradation and perception have been common engineering targets for enhanced shelf life and freshness of climacteric fruits (Klee and Giovannoni, 2011). Suppression of biosynthesis genes in the ethylene pathway such as aminocyclopropane-1-carboxylic acid (ACC) synthase (Oeller et al., 1991) or ACC oxidase (Hamilton et al., 1990) became a favoured strategy for researchers. ACC, the critical intermediate in ethylene biosynthesis was also made limiting by respective overexpression of S-adenosyl methionine hydrolase (Good et al., 1994) or ACC deaminase (Klee et al., 1991). Cell wall architecture, a major determinant of fruit texture, is governed by more than 50 genes (Sato et al., 2012; The Tomato Genome Consortium, 2012). Lowered ethylene delays fruit softening, characteristic of ripening, by hindering textural changes induced by cell wall degradation enzymes such as pectin methylesterase (PME) and PG. PG was downregulated by constitutive expression of chimeric anti-PG in tomato cultivar Ailsa Craig (Smith et al., 1990). Transgenic line homozygous for anti-PG in the T2 generation had 99% reduction in PG activity and concomitant decrease in pectin depolymerization. Ethylene and lycopene accumulation were not affected by anti-PG expression. PME, believed to have a role in determining fruit texture, was also downregulated by anti-PME3 driven by CaMV 35S promoter in tomato. No effect on fruit softening was observed in these fruits, but fortuitously juice viscosity and total soluble solids were enhanced (Tieman et al., 1992; Tieman and Handa, 1994; Thakur et al., 1996a,b). Suppression of ripening-specific N-glycoprotein modifying enzymes, α-mannosidase and β-D-N-acetylhexosaminidase, led to firmer fruits relative to controls and enhanced shelf life of tomatoes (Meli et al., 2010).

Chlorophyll degradation and accumulation of carotenoids are hallmarks of ripening tomato (Alexander and Grierson, 2002). Phytoene synthase (Psy) catalyses the first step in carotenoid biosynthesis pathway, the condensation of two molecules of geranyl geranyl diphosphate (GGPP) into phytoene. Overexpression of *Psy1* gene driven by CaMV *35S* promoter in tomato resulted in high lycopene content but with dwarf plants that exhibited 30-fold decrease in gibberellin GA_1 (Fray et al., 1995). It was postulated that lycopene synthesis occurred at the expense of gibberellins by diverting the common GGPP precursor to carotenoid pathway because of the constitutive *Psy1* overexpression in the transgenic tomato. Mitigation of the dwarfing phenotype was achieved when fruit-specific PG promoter was used to drive the expression of *crtB* (*Psy* from *Erwinia uredovora*) (Fraser et al., 2002). The bacterial gene was fused with tomato *Psy1* transit sequence for chromoplast targeting. Suppression by RNAi of *DET1* (DEETIOLATED 1), a negative regulator of photomorphogenesis, increased the content of both carotenoids and flavonoids with no negative effects on the plant phenotype of the transgenic tomatoes (Davuluri et al., 2005).

Flavonoids are polyphenolic hydrophilic, aromatic small molecules synthesized via the phenylpropanoid pathway (Ververidis et al., 2007). Malonyl-CoA and *p*-coumaroyl-CoA, derived from carbohydrate metabolism and phenylpropanoid pathway, respectively, are the reactant molecules for most flavonoids (Forkmann and Heller, 1999). These compounds, abundant in fruit cuticles (Hunt and Baker, 1980), contribute to fruit colour, flavour and texture. Tomato pericarp is deficient in flavonoids due to low or no expression of the biosynthesis genes (Bovy et al., 2007). Metabolic engineering strategies for flavonoid compounds in tomato include (i) overexpression of structural and regulatory heterologous genes of flavonoid pathway, (ii) RNA interference for blocking steps/branches of the pathway and (iii) introducing novel pathway branches for new flavonoids (Bovy et al., 2007). A four-gene construct containing *Petunia* chalcone synthase (*CHS*), chalcone isomerase (*CHI*), flavanone-3-hydroxylase (*F3H*) and flavonol synthase (*FLS*) was used in an attempt to upregulate levels of flavonols in fleshy tomato fruit (Colliver et al., 2002). Orchestrated action of all four genes led to increased levels of flavonols in both peel (quercetin glycosides) and flesh (kaempferol glycosides). Fruit-specific expression using the E8 promoter in front of two transcription factors, *Rosea1* (*Ros1*) and *Delila* (*Del*), which are activators of anthocyanin biosynthesis enhanced anthocyanin production in tomato pericarp at concentrations comparable to that in blackberries and blueberries (Butelli et al., 2008). Identification and quantification of seven anthocyanins by LC-MS/MS including two novel anthocyanins, malvidin-3-(*p*-coumaroyl)-rutinoside-5-glucoside and malvidin-3-(feruloyl)-rutinoside-5-glucoside, was carried out in transgenic purple tomatoes expressing transcription factors, *Del* and *Ros1* (Su et al., 2016).

RNA interference-mediated silencing of tomato *CHS1* led to a significant reduction (99% reduction of total flavonoids) of naringenin chalcone and quercetin rutinoside in comparison with wild-type controls (Schijlen et al., 2007). Such silenced tomato fruits had abnormal phenotype and showed parthenocarpic fruit development. Similarly, seedless fruits or ones with reduced seed set were obtained when grape stilbene synthase was expressed (Giovinazzo et al., 2005; Schijlen et al., 2006) in tomato. These observations indicated an important role of flavonoids in fertilization, seed and fruit development. Fruit-specific expression of the *Arabidopsis MYB12* transcription factor in tomato led to production of novel bioactive flavonoids, particularly flavonols and caffeoyl quinic acids (Zhang et al., 2015). ChIP-qPCR (chromatin immunoprecipitation-quantitative PCR) assays showed that *AtMYB12* bound directly to promoters of genes involved in both primary and secondary metabolism. Co-expression of feedback-insensitive *Escherichia coli* 3-deoxy-D-arabinoheptulosonate 7-phosphate synthase (*AroG*) and petunia MYB transcript factor, *ODORANT1* (*ODO1*), in tomato fruits led to a dual effect on phenylalanine and related biosynthetic pathways (Xie et al., 2016). Positive impact was seen on the content of tyrosine and metabolites derived from coumaric and ferulic acids but secondary metabolites downstream the phenylalanine pathway, including kaempferol, naringenin and quercitin-derived metabolites, as well as aromatic volatiles were negatively impacted.

4 Abiotic stress tolerance

Plants have developed very sophisticated molecular mechanisms to evade stress situations, likely because of being sessile. Environmental extremes therefore impact plant growth and development on a daily basis, and in wake of the current global climate change additional unfavourable effects on plant yield is furthered by abiotic stresses including salinity, drought, high light, flooding, wounding, cold and heat (Ainsworth and McGrath, 2010). Central to such negative impact involves overlapping cell signalling controls and molecular mechanisms (Zhu, 2002; Pandey et al., 2011; Pandey, 2015). Table 2 summarizes various efforts on elucidating mechanisms and molecular manipulation of tomato for tolerance to abiotic stresses, some of which are further discussed below.

Table 2 Genetic engineering of tomato for tolerance to abiotic stress

Gene and source	Description	Trait/phenotype conferred	Reference
D-Galacturonic acid reductase *GalUR* *Fragaria* x *ananassa* Strawberry	Overexpression of *GalUR* driven by CaMV *35S* (Cauliflower mosaic virus *35S* promoter)	Abiotic stress tolerance, particularly salt tolerance (200 mM), elevated ascorbic acid content	Lim et al., 2016
SOS2L1 Salt overly sensitive *Malus x domestica* Apple	Overexpression of full-length *MdSOS2L1* cDNA driven by CaMV *35S*	Salt tolerance (300 mM NaCl)	Hu et al.,2016
WD6 (family of Trp-Asp WD-repeat proteins) *Solanum lycopersicum*	Constitutive expression by *Agrobacterium*-mediated transformation	Drought and salt tolerance	Yang et al., 2015
NHX2 Na$^+$/H$^+$ Exchanger *S. lycopersicum*	*LeNHX2* ion transporter overexpression driven by CaMV *35S*	Salt tolerance (120 mM NaCl), higher Na$^+$/H$^+$ and K$^+$/H$^+$ transport activity in root intracellular membrane vesicles, two fold higher K$^+$ depletion rate, half cytosolic K$^+$ activity, enhanced HAK (high-affinity K$^+$ uptake system) expression under K$^+$-limiting conditions	Huertas et al., 2013

Table 2 (*Continued*)

Gene and source	Description	Trait/phenotype conferred	Reference
BADH Betaine aldehyde dehydrogenase *Suaeda liaotungensis*	Overexpression of *BADH* gene driven by CaMV *35S* and P5 promoters	Salt tolerance (200 mM NaCl)	Wang et al., 2013
GlyI and *GlyII* Glyoxalase I and glyoxylase II *Brassica juncea* and *Pennisetum glaucum*	Overexpression of *BjGlyI* and *PgGlyII* genes under CaMV *35S*	Engineering of glyoxyalase detoxification provides salt tolerance (800 mM NaCl)	Álvarez-Viveros et al., 2013
SOS2 *S. lycopersicum*	*SlSOS2* ion transporter overexpression driven by CaMV *35S*	Salt tolerance (120 mM NaCl), higher Na$^+$ content in leaves and stems, no differences in K$^+$ content relative to untransformed plants	Huertas et al., 2012
Hem1 *Saccharomyces cerevisiae*	*Hem1* gene driven by the light-responsive HemA1 promoter from *Arabidopsis thaliana*	Improved salt tolerance (200 mM NaCl)	Li et al., 2012
Na$^+$/H$^+$ antiporter *NHX* *P. glaucum* Vacuolar H$^+$-pyrophosphatase *AVP1* *Arabidopsis thaliana*	Co-expression of *PgNHX1* and *AVP1* driven by CaMV *35S*	Tolerance to 200 mM NaCl, accumulated proline and Na$^+$, 1.4 and 1.5 times, respectively, than single gene transformants	Bhaskaran and Savithramma, 2011
Xyloglucan endo-*trans*-gluco-sylase/hydrolase (*XTH3*) *Capsicum annum*	Ectopic overexpression of full-length *CaXTH3* cDNA driven by CaMV *35S*	Increased tolerance to salt (100 mM) and drought stresses	Choi et al., 2011
codA Choline oxidase *Arthrobacter globiformis*	Expression of *codA* gene fused with chloroplast targeting transit peptide driven by CaMV *35S*	Accumulation of glycinebetaine in leaves up to 297 nmol g^{-1} fresh weight, provides salt (200 mM NaCl) and water stresses	Goel et al., 2011
SOS1 *S. lycopersicum*	Silencing of *SlSOS1* transporter by RNAi technology	Transgenic plants with reduced expression of *SlSOS1* showed reduced growth relative wild type in saline conditions. They accumulated higher Na$^+$ in leaves and roots than stems under salt stress	Olías et al., 2009

(*Continued*)

Table 2 (*Continued*)

Gene and source	Description	Trait/phenotype conferred	Reference
TPS1 Trehalose-6-phosphate synthase *S. cerevisiae*	Overexpression of *TPS1* gene driven by CaMV *35S*	Increased tolerance to drought, salt and oxidative stress; abnormal plant morphology	Cortina and Culianez-Macia, 2005
BADH Betaine aldehyde dehydrogenase *Atriplex hortensis*	Overexpression of *BADH* gene driven by CaMV *35S*	Salt tolerance (120 mM NaCl)	Jia et al., 2002
NHX1 *A. thaliana*	Overexpression of *At-NHX1* gene driven by CaMV *35S*	Salt tolerance (up to 200 mM NaCl), high Na^+ accumulation in leaves, very low levels in fruits, increased growth, flower and seed production	Zhang and Blumwald, 2001
betA Choline dehydrogenase *Escherichia coli* TG1	Overexpression of *beta* gene under the control of CaMV *35S*	Increased osmotic adjustment ability of transgenics relative to wild type, salt tolerance at 200 mM NaCl	Wang et al., 2001
HAL1 Halotolerance gene *S. cerevisiae*	Overexpression of *HAL1* open reading frame under CaMV *35S*	Modulate cation transport systems, Na^+ and K^+ homeostasis, reduced growth, fruit yield reduction in normal conditions but 27% increase under salt stress	Gisbert et al., 2000 (short-term study); Rus et al., 2001 (long term study)
BADH-1 Betaine aldehyde dehydrogenase *Sorghum*	Production of hairy roots by transformation with Ri plasmid	Maintenance of osmotic potential under salt stress	Moghaieb et al., 2000
HAL2 Halotolerance gene *S. cerevisiae*	Overexpression of *HAL2* gene driven by CaMV *35S* with double enhancer and synthetic Alfalfa mosaic virus RNA4 leader sequence	Salt tolerance (175 mM NaCl)	Arrillaga et al., 1998
ICE1 [Inducer of CBF (C-repeat binding factor expression)] *A. thaliana*	Overexpression of *ICE1*	Tolerance to low temperature stress, high proline, peroxide content and catalase activity in transgenics but low malondialdehyde content	Juan et al., 2015
CRT/DRE-binding factor1 *A. thaliana*	Expression of *AtCBF1* driven by stress-inducible ABRC1 promoter from barley *HAV22* gene	Enhanced tolerance to chilling, water deficit and salt stress	Lee et al., 2003a

Table 2 (*Continued*)

Gene and source	Description	Trait/phenotype conferred	Reference
SAMDC S-adenosylmethionine decarboxylase *S. cerevisiae*	*Agrobacterium*-mediated leaf disc transformation	Accumulation of spermidine and spermine under high temperature stress, high antioxidant enzyme activity, protection of membranes from lipid peroxidation	Cheng et al., 2009
Osmotin Tobacco	Overexpression of *osmotin* gene driven by CaMV *35S*	Transgenic lines showed higher relative water content, chlorophyll content, proline content and leaf expansion than wild type under salt and drought stress	Goel et al., 2010
OaAANAT Arylalkylamine *N*-acetyltransferase *OaHIOMT* Hydroxyindole *O*-methyltransferase *Ovis aries* (sheep)	Expression of *OaAANAT* and *OaHIOMT* genes driven by constitutive CaMV *35S*	Drought tolerance in transgenic lines expressing *OaHIOMT*	Wang et al., 2014

4.1 Salinity stress tolerance

Salinity negatively affects productivity and quality of cultivated crop plants. It is anticipated that salinization of arable land will increase from 20% presently to 50% by the year 2050 (Wang et al., 2003), perhaps due to increased use of underground water, lack of freshwater and flawed fertilization practices (Yu et al., 2012). Salt tolerance is a complex trait and response to salt stress is attuned by agronomic, physiological, developmental, genetic and environmental factors (Foolad, 2004). Studies on mitigating salinity stress in plants have involved following strategies.

4.1.1 Ion transport proteins

Ion transport proteins integral to plant plasma and tonoplast membranes are important for maintenance of intrinsic ion balance in a cell. Transporters that maintain Na$^+$ balance in plants include NHX (Na$^+$/H$^+$ exchanger), SOS (salt overly sensitive) and HKT (high-affinity potassium transporter).

Roots of T3 homozygous tomato plants with a single insertion of *LeNHX2* under CaMV *35S* promoter, an endosomal class II NHX transporter, accumulated *LeNHX2* transcripts under stress by 120 mM salt. These transgenic plants yielded higher shoot fresh weight under salt treatment; however, under K$^+$-limiting conditions shoot fresh weight decreased relative to the wild-type control plants. When external K$^+$ availability is low, the decrease of cytosolic K$^+$ caused by *LeNHX2* overexpression could lead to the higher LeHAK5 (high-affinity K$^+$ uptake system) expression in transgenic plants relative to untransformed control plants. Overexpression of *LeNHX2* resulted in increased uptake of K^+ by epidermal root cells. Plant growth is inhibited under K^+ deficiency and salt tolerance is imparted due to modified K^+ uptake (Huertas et al., 2013). Overexpression of *Arabidopsis* vacuolar Na^+/H^+ antiport (*AtNHX1*) enabled transgenic tomato plants to grow, flower and produce fruit in the presence of 200 mM NaCl (Zhang and Blumwald, 2001). Notably, high amounts of sodium accumulated in leaves but not the fruit.

Activation of Ca^{2+}-dependent SOS signalling is a key molecular mechanism to prevent intracellular accumulation of toxic amounts of Na^+ (Zhu, 2002) with *SOS1*, *SOS2* and *SOS3* as the functional components (Zhu et al., 1998). Compartmentalization of Na^+ in plant stems as opposed to roots and leaves is a common mechanism used by SOS system to help plants evade effects of high salt stress. Decreasing *SlSOS1* transcript population in transgenic tomato by RNAi technology changed organ partitioning of Na^+ with more accumulation in leaves and roots and lesser in the stem (Olías et al., 2009) confirming the function of SOS1. Ectopic expression of *MdSOS2L1*, a CIPK protein kinase (calcineurin B-like protein-interacting protein kinase), in tomato enabled tolerance to NaCl at 300 mM concentration (Hu et al., 2016). These transgenic tomatoes contained higher levels of procyanidin and malate, and produced less reactive oxygen species (ROS) such as hydrogen peroxide that may have contributed to containing oxidative damage. Overexpression of *SlSOS2*, a candidate gene believed to be involved in regulating vacuolar Na^+/H^+ exchange, in tomato conferred salinity tolerance (at 120 mM NaCl) via Na^+ extrusion from the root, accumulation in aerial parts, active loading of Na^+ into xylem, and compartmentalization of Na^+ and K^+ (Huertas et al., 2012). These transgenic tomato plants also had higher transcript levels of the Na^+, K^+/H^+ antiporter *LeNHX4* in roots, stems and leaves, and higher Na^+/H^+ antiport activity in root tonoplast vesicles.

4.1.2 Osmotic homeostasis

Whether higher plants possess an osmoregulatory mechanism to cope up with water stress was reviewed (Schobert, 1977) and it was proposed that polyols and proline play a key role in osmotic adjustments. Mannitol, a sugar alcohol synthesized from fructose by mannitol-1-phosphate dehydrogenase (*mtlD*), plays an important role in the osmoregulation of plants. Thus, transgenic tomato plants that constitutively expressed *E. coli mtlD* gene were found to develop tolerance to chilling, drought and salinity stress (Khare et al., 2010).

Osmotin or osmotin-like protein is a member of the PR-5 family (class 5 pathogenesis-related) induced in plants upon exposure to abiotic and biotic stresses and implicated in providing osmotolerance (Singh et al., 1985, 1987; Barthakur et al., 2001). It is constitutively present in tomato grown under sustainable agricultural practices and provides tolerance against biotic stress (Kumar et al., 2004, 2005). Overexpression of tobacco osmotin gene in tomato conferred increased tolerance to salt and drought stress (Goel et al., 2010) in these transgenics. These plants had higher relative water content, chlorophyll, proline and leaf expansion than the wild-type plants when exposed to stress. Osmotin provides protection to native proteins during stress and repairs denatured proteins (Pandey et al., 2011). Overexpression of tobacco osmotin gene is also believed to modulate transcriptional profiles of other pathogenesis-related (PR) genes such as PR1 and PR4. These PR genes may provide resistance to pathogens since they are increased in tomato seedlings treated with exogenous salicylic acid and benzo (1,2,3) thiodiazole-7-carbothioic acid *S*-methyl ester (Fiocchetti et al., 2006).

4.1.3 Production of ROS scavengers

Oxidative stress in plant cells upon exposure to high saline conditions results in increased production of ROS (Mittler et al., 2004) and subsequent oxidative damage, characterized by lipid peroxidation and fatty acid de-esterification, in cell membranes (Arora et al., 2002). Genes imparting salt tolerance seem to be present in both halophytes and glycophytes; however, they are regulated differently in varied environments (Himabindu et al., 2016). Salt-responsive genes from halophytes have been engineered in various plant systems, including tomato (Himabindu et al., 2016). Notably, expression of genes providing salt tolerance is advantageous when transgenes are fused to stress-inducible promoters as the negative effects on the transgenic plants are prevented in the process. For instance, expression of *BADH* (betaine aldehyde dehydrogenase) gene driven by P5 promoter (from *Suaeda liaotungensis*) was used to develop tomato transgenics that were tolerant to sodium chloride at 200 mM (Wang et al., 2013).

Transgenic tomatoes expressing strawberry *GalUR* (D-galacturonic acid reductase) gene were more tolerant to abiotic stresses induced by methyl viologen (20 µM), NaCl (200 mM) and mannitol (up to 300 mM) than the control plants (Lim et al., 2016). These transgenic lines contained higher ascorbic acid and chlorophyll contents and low levels of malondialdehyde under salt stress. Higher expression levels of antioxidant genes such as ascorbate peroxidase (*AOX*) and catalase (*CAT*) were also found in the transgenic plants compared to controls.

4.2 Chilling stress tolerance

Tomato is cold sensitive and suffers chilling injury below 13°C. At such cold temperatures the plant growth and yield are compromised (Lin et al., 2000). The identity of cold-induced genes have come to light in numerous works of researchers, implicating diverse players, including transcription factors such as CBF (CRT/DRE-binding factor), the disaccharide trehalose and reduced glutathione (GSH) (Mackenzie et al., 1988; Thomashow, 1999; Herbette et al., 2005; Li et al., 2010). A composite plant defence response to chilling stress involves the regulatory molecular cascade, which includes cold-responsive genes/transcription factors ICE1 (inducer of CBF3 expression1), MYB, MYC and CBF (C-repeat binding factor), along with ubiquitin E3 ligase HOS1 and SUMO E3 ligases SIZ1/SIZ2 (Thomashow, 2010; Medina et al., 2011; Knight and Knight, 2012). Low-temperature-responsive model for *Arabidopsis* involves ICE1 that encodes a MYC-like basic helix–loop–helix transcription factor that, in turn, activates CBF3/DREB1A and COR genes (Chinnusamy et al., 2003).

The plant expression construct p3301 harbouring *Arabidopsis ICE1* transcription factor was overexpressed in tomato cultivar rhubarb with the aim of achieving cold tolerance (Juan et al., 2015). The transgenics were given a low-temperature stress that led to an increase in peroxidase and catalase activities along with proline content, with the oxidative stress molecule malondialdehyde being at lower concentrations than the control. Transgenic tomato cultivar Pusa Ruby transformed to express tobacco osmotin gene was subjected to cold temperature treatment (4°C for 2 and 24 h) and it led to higher expression of additional stress-responsive genes, namely, CBF1, P5CS, and APX and increased accumulation of free proline and ascorbic acid content (Patade et al., 2013).

A revealing finding upon analysis of ectopically expressed ICE1, which led to enhanced tolerance to cold stress, was the concomitant increase in arginine decarboxylase (ADC) transcripts and levels of free polyamines (Huang et al., 2015). Polyamines are biogenic amines that impart protection to plants against different abiotic stresses (Mattoo, 2014; Mattoo et al., 2015). Interestingly, while studying chilling temperature responses of tomato fruit, a PR protein, PR1b1 protein, was found to predominantly accumulate in fruit brought to room temperature after two weeks of exposure at 2°C. In tomato lines engineered for the ripening-associated accumulation of higher polyamines, spermidine and spermine, the PR1b1 protein remained abundant in re-warmed chilled fruit for an extended period as compared to control fruit deficient in these two polyamines (Goyal et al., 2016). Further, a positive correlation was found between increase in the PR1b1 protein and gene transcripts with the transcripts of MYC2, MYB1, CBF1 and glucose-6-phosphate dehydrogenase (G-6-P DH) transcripts, and salicylic acid levels in the high spermidine/spermine transgenic tomato fruit (Goyal et al., 2016). It was proposed that polyamine-mediated accumulation of PR1b1 protein in re-warmed chilled tomato could be a pre-emptive plant defence mechanism related to cold stress-induced disease resistance (SIDR) phenomenon, function and mechanisms of which are yet to be determined (Moyer et al., 2015).

Trehalose, an osmoprotectant, is yet another mitigator of chilling stress through protection of membrane proteins, stabilization of native protein state, reduction in aggregation of denatured proteins and prevention of cell desiccation (Crowe et al., 1984; Singer and Lindquist, 1998). Thus, rice TPP1 (trehalose-6-phosphate phosphatase 1) overexpression augmented the cold tolerance of rice (Pramanik and Imai, 2005) while yeast TPS1 (trehalose-6-phosphate synthase 1) expression in tomato imparted drought tolerance (Cortina and Culianez-Macia, 2005). Similarly, overexpression of mouse GSH peroxidase in transgenic tomato plants led them to retain normal photosynthetic activity under chilling stress compared to wild-type plants (Herbette et al., 2005).

4.3 Drought tolerance

Plants that grow in arid conditions have evolved a highly expanded root system, adaptations like spines to reduce transpiration and waxy cuticles on leaves to evade scarcity of water (Kramer and Boyer, 1995). Arabidopsis CBF1 overexpressed in tomato imparted resistance to drought stress (Hseih et al., 2002). A slight increase in water stress tolerance was observed in transgenic tomato expressing a highly heat-stable Populus tremula 66-kD protein (Roy et al., 2006). Arabidopsis vacuolar H+-pyrophosphate (AVP1) expression in tomato that resulted in salt- and drought-tolerant phenotype was accompanied with an increase in root biomass (Park et al., 2005). Relatively more drought-tolerant transgenics than the wild-type controls were obtained when Saccharomyces cerevisiae TPS1 gene was introduced into tomato, albeit with abnormal changes in plant morphology (Cortina and Culianez-Macia, 2005). Constitutive expression of sheep arylalkylamine N-acetyltransferase in tomato imparted drought tolerance to the transgenic lines (Wang et al., 2014).

4.4 High temperature tolerance

Chaperones or heat shock proteins are well-known components of plant machinery to respond to high temperature stress (Swindell et al., 2007). Tomato fruits transformed with Arabidopsis heat shock factor (hsf) gene were tolerant equally to high (47°C) and low (−2°C) temperatures during storage periods of up to 4 weeks (Lurie et al., 2003). Overexpression of *Arabidopsis* ERECTA (*ER*) in transgenic tomato and rice lines improved their tolerance to high temperature, both in greenhouse and field tests at multiple locations in China, over several seasons, and had increased biomass (Shen et al., 2015). These researchers used *S. pimpinellifolium* (accession LA1589), which is temperature sensitive, compared with modern-day varieties of tomato to overexpress ER under CaMV35 promoter. ER over-expression in tomato resulted in two times larger leaf size, decreased stomatal density, decreased stomatal conductance and increased transpiration efficiency relative to wild type. The transgenic tomatoes could withstand temperature regime of 40°C/28°C (day/night) for 10 days in a growth chamber.

As mentioned above, polyamines are important regulators in plant biology and defence providers. They regulate multiple biological processes in plants, including stomatal opening, stress responses and interaction with other plant hormones (Kumar et al., 2006; Liu et al., 2007; Yang et al., 2007; Handa and Mattoo, 2010; Anwar et al., 2015). Overexpression of yeast S-adenosylmethionine decarboxylase in tomato boosts endogenous concentrations of polyamines, and such transgenic plants have been shown to have superior tolerance to high temperature stress than the control plants (Cheng et al., 2009).

5 Biotic stress tolerance

Biotic stress due to bacterial, viral, fungal and insect pathogens devastates the yield and quality of crop plants, including tomatoes. Molecular markers have provided fundamental data regarding pathogen population diversity and evolution important for disease control in different crop plants (De Giovanni et al., 2004; Kaur et al., 2005; Purkayastha et al., 2006, 2008). In tomato, 40 widely spread diseases are known, most of which are caused by either bacteria or fungi (Khaliluev and Shpakovskii, 2013). Pathogen recognition is followed by a rapid oxidative burst typified by production of reactive oxygen intermediates such as superoxide anion (O_2^-), hydroxyl radicals (OH) and hydrogen peroxide (H_2O_2), which may control pathogen resistance response and acquired immunity (Jabs, 1999; Grant and Loake, 2000; Bergougnoux, 2014). The response to pathogen attack is mediated by active and passive defences including hypersensitive reaction, programmed cell death/apoptosis, defence genes expression (PR proteins), phytoalexins, phytoanticipins, phenolics and ROS. Table 3 lists selected examples of genetic engineering of tomato to mitigate biotic stresses.

5.1 Resistance to bacterial diseases

Overexpression of potato polyphenol oxidase (*StPPO*) gene in tomato imparted resistance to *Pseudomonas syringae* pv. *tomato*, causal agent of bacterial spot disease, in terms of both number and area of lesions corresponding to a 100-fold reduction in bacterial population in infected leaves (Li and Steffens, 2002). These plants accumulated quinones that had a cytotoxic effect on pathogens. In contrast, suppression of PPO gene in tomato by antisense *StPPO* cDNA dramatically decreased the oxidation of caffeic acid and increased the plant's susceptibility to this pathogen (Thipyapong et al., 2004), suggesting the involvement of phenolic oxidation in disease resistance in tomato. The transgenic tomato plants expressing a human cathelicidin antimicrobial peptide (hCAP18/LL-37)

Table 3 Genetic engineering of tomato for tolerance to biotic stress

Gene and source	Description	Trait/phenotype conferred	Reference
Cathelicidin antimicrobial peptide *Homo sapiens*	Overexpression of *hCAP18/LL-37* (fused to signal peptide from pea *vc-2* gene) driven by PGD1 promoter, extracellular localization of mature protein	Resistance to bacterial soft rot and bacterial spot diseases, high expression of PR protein, LTP (protease inhibitor contains lipid transfer protein) and AFP1 (cysteine-rich antifungal protein precursor) genes	Jung, 2013
MAP kinase	Silencing of *SlMAPKKKε*	Disruption of resistance against *Xanthomonas campestris* and *Pseudomonas syringae*	Melech-Bonfil and Sessa, 2010
Cys-2/His-2-zinc finger protein-TF (pathogenesis-induced factor) *Capsicum anum*	Overexpression of *CaPIF1* driven by CaMV *35S*	Resistance to *Pseudomonas syringae* pv. *tomato* DC 3000, tolerance to cold stress	Seong et al., 2007
Polyphenol oxidase *Solanum tuberosum*	Suppression of *StPPO* by antisense technology driven by CaMV *35S*	Increased susceptibility to *P. syringae* pv. *tomato*	Thipyapong et al., 2004
Magainin-cationic antimicrobial peptide Synthetic	Overexpression of *MSI-99* (fused to signal peptide from pea *vicilin* gene) driven by EnhancerCaMV *35S*, targeted expression in extracellular spaces	Increased resistance to *P. syringae* pv. *tomato* (bacterial speck pathogen), no cytotoxic effects in transgenic plants	Alan et al., 2004
Glycoprotein, antibacterial *H. sapiens*	Lactoferrin (LF)	Partial resistance to *Ralstonia solanacearum* (bacterial wilt)	Lee et al., 2002
Polyphenol oxidase *S. tuberosum*	Overexpression of *StPPO* driven by CaMV *35S*	Increased resistance to *P. syringae* pv. *tomato*	Li and Steffens, 2002
Serine/threonine protein kinase (*R* gene) *S. lycopersicum*	Overexpression of *SlPTO* driven by CaMV *35S*	Resistance to *Xanthomonas campestris* pv. *vesicatoria* and *Cladosporium fulvum*	Tang et al., 1999

Table 3 (*Continued*)

Gene and source	Description	Trait/phenotype conferred	Reference
Hevein-like protein (PR4 family) *Pharbitis nil*	Overexpression of *Pn-AMP2* driven by CaMV *35S*	Enhanced resistance to *Phytophthora capsici* and *Fusarium oxysporum*	Lee et al., 2003b
Bt Cry2A *Bacillus thuringiensis*	Overexpression of Bt Cry2A gene driven by CaMV *35S*	Resistance to neonate larvae of *Helicoverpa armigera* in laboratory conditions	Hanur et al., 2015
Chymotrypsin inhibitor *S. lycopersicum*	Overexpression of JIP21 (jasmonic-induced protein) gene driven by CaMV *35S*	Increased mortality of lepidopteran *Spodoptera littoralis* larvae	Lison et al., 2006
Chitinase *Win6* Poplar	Potato virus X CP (coat protein) promoter: *win6*	Resistance to Colorado potato beetle (*Leptinotarsa decemlineata*) larvae	Lawrence and Novak, 2006
Arginase *S. lycopersicum*	Overexpression of *ARG2* gene driven by CaMV *35S*	Increased resistance to *Manduca sexta* larvae	Chen et al., 2005
Pin and carboxypeptidase inhibitors *S. tuberosum*	Tissue-specific expression of serine protease inhibitor, PI-II driven by *StLS1* promoter (leaf and stem specific) and carboxypeptidase inhibitor (PCI) driven by *rbsc-1A*	Increased resistance to *Heliothis obsoleta* and *Liriomyza trifolii*	Abdeen et al., 2005
Bt Cry1Ac *B. thuringiensis*	Overexpression of *Bt Cry1Ac* gene driven by CaMV *35S*	Resistance to larvae of *Helicoverpa armigera* in leaves and fruits	Mandaokar et al., 2000
δ-Endotoxin gene *B. thuringiensis* subsp. *tenebrionis*	Overexpression of Bt toxin gene driven by CaMV *35S*	Resistance to Colorado potato beetle (*Leptinotarsa decemlineata*) larvae	Rhim et al., 1995
Systemin *S. lycopersicum*	Silencing prosystemin gene driven by CaMV *35S* by antisense technology	Decrease in resistance to *Manduca sexta* (tobacco hornworm) larvae via reduction in proteinase inhibitors I and II	Orozco-cardenas et al., 1993

were significantly resistant to bacterial soft rot and bacterial spot with concomitant strong expression of PR protein, LTP and AFP1 genes. Transgenic tomato leaf protein extracts limited the growth of *P. carotovorum* ssp. *carotovorum* to 15%, and that of *Xanthomonas campestris* pv. *vesicatoria* to 35% (Jung, 2013). MAP (mitogen-activated protein) kinase signalling pathways are associated with plant immunity with their involvement in hypersensitive response cell death and resistance against Gram-negative bacterial pathogens in tomato. Silencing of *SlMAPKKKε* in tomato disrupted resistance against *X. campestris* and *P. syringae* (Melech-Bonfil and Sessa, 2010). Downregulation of a peroxidase gene *Ep5C* by its antisense RNA imparted resistance to *P. syringae* pv. *tomato* (Coego et al., 2005).

5.2 Resistance to fungal and viral diseases

PR proteins and antimicrobial peptides are effective at micromolar concentrations (and non-toxic to animals and humans) by imparting resistance to fungal (and bacterial) diseases in plants (Khaliluev and Shpakovskii, 2013). *Agrobacterium*-mediated transformation of tomato cultivar A53 with dual PR genes, tobacco *AP24* osmotin and bean chitinase, produce transformants with improved *Fusarium* wilt resistance (Ouyang et al., 2005). Resistance genes (*R* genes) provide plants the tools for pathogen effector recognition and, therefore, race-specific immunity. Transgenic tomato expressing *S*-receptor-like kinase (SRLK) genes I, I-2 and I-3 imparted tolerance to *F. oxysporum* f. sp. *lycopersici* races 1, 2 and 3, respectively (Catanzariti et al., 2015). A 36% to 58% reduction of *Fusarium* wilt was demonstrated when tobacco class I chitinase and 1,3-glucanase genes were introduced into tomato (Jongedijk et al., 1995).

Agrobacterium-mediated transformation of tomato with acidic endochitinase (*pcht28*) from wild *S. chilense* developed resistance against *Verticillium dahliae* (races 1 and 2) tested in greenhouse (Tabaeizadeh et al., 1999). Also, transgenic tomato fruits expressing hevein (*HEV1*) were less prone to the fungal pathogen *Trichoderma hamatum* (Lee and Raikhel, 1995), while hevein-like chitin-binding protein PnAMP2 from *Pharbitis nil* conferred resistance to *Phytophthora capsici* in transgenic tomato (Lee et al., 2003b). Chitin-binding protein genes from *Amaranthus caudatus* (*ac*) and hevein-like antimicrobial protein from *Stellaria media* (*amp1, amp2*) in transgenic tomato plants augmented their resistance to late blight.

In a relatively uncommon instance of genetic engineering, expression in tomato of a single gene involved in ergosterol biosynthesis, C-5 sterol desaturase (*FvC5SD*), from *Flammulina velutipes* elevated protection against *Sclerotinia sclerotiorum* through a thicker waxy cuticle barrier to entry (Kamthan et al., 2012). In addition, it also enabled drought tolerance, an increase in iron as well as polyunsaturated fatty acid content in tomato. The inhibitor-of-virus replication (*IVR*) gene from tobacco introduced into tomato cultivar VF36 conferred partial resistance to several fungal pathogens, namely, *Alternaria alternata*, *Pythium aphanidermatum* and *Rhizoctonia solani* at seedling stage and to *A. solani* (early blight) and *Oidium neolycopersici* (powdery mildew) in mature plants (Elad et al., 2012).

Resistance to tomato mosaic virus (TMV) and to high temperature was engineered into tomato by cloning *Arabidopsis NPR1* gene (Lin et al., 2004). In addition to the resistance to TMV, these transgenics also displayed resistance to *Fusarium* wilt, grey leaf spot (fungal diseases), bacterial wilt and bacterial spot. Transgenic tomatoes expressing intron-hairpin construct derived from *C1* gene enabled post-transcriptional silencing of tomato yellow leaf curl virus (Fuentes et al., 2006). Overexpression of the *SlAOX* gene, encoding a tomato mitochondrial alternative oxidase (AOX), enhanced tolerance to spotted wilt virus in tomato (Ma et al., 2011).

5.3 Resistance to insects and nematodes

One of the early reports of field testing of transgenic tomato plants was the one expressing insecticidal protein from *Bacillus thuringiensis* var. *kurstaki* HD-1 specific against lepidopterans (Delannay et al., 1989). These transgenic plants were resistant to leaf damage by *Manduca sexta*, *Heliothis zea* and *Keiferia lycopersicella*. Plant defence systems are equipped with proteinase inhibitors and secondary metabolites for protection against insects. Tomato leaves contain systemin, an 18 amino acid polypeptide, which induces proteinase inhibitors, and is a systemic wound signal, likely mediates jasmonate signalling pathway in response to an insect attack. Plant proteinase inhibitors inhibit the activity of gut insect proteases. Proteases in the gut of insects break down proteins to produce amino acids as food. Examples include lepidopteran serine proteinases and coleopteran cysteine and aspartic proteinases. To manage defence against insects, transgenic tomato plants were developed that expressed two potato protease inhibitors (Abdeen et al., 2005). A serine proteinase inhibitor, PI-II, and a carboxypeptidase inhibitor, PCI, in combination, provided strong resistance against *Heliothis obsoleta* and *Liriomyza trifolii* larvae, respectively, in homozygous transgenics. The South Indian tomato cultivar, Arka Vikas, was transformed using *Agrobacterium* carrying a *BtCry2A* construct to create transgenics resistant to damage caused by neonate larvae of *Helicoverpa armigera* (Hanur et al., 2015).

Root-knot nematode (RKN) pathogen, *Meloidogyne incognita*, causes major economic losses in agriculture. Reducing *expansin* gene expression in tomato gall cells by antisense *LeEXPA5* (tomato expansin isoform; expansin precursor 5 locus) was found to limit pathogenesis by RKN, which was attributed to inability of the nematode to complete its life cycle (Gal et al., 2006). RKN control is also possible by plant defence genes, for instance, proteinase inhibitors. Cysteine proteinase inhibitor *CeCPI* from *Colocasia esculenta* and fungal chitinase *PjCHI-1* from *Paecilomyces javanicus* were overexpressed together in tomato, and the transgenic plants had an unfavourable effect on chitin content of RKN eggs as well as on embryogenesis (Chan et al., 2015).

6 Tomato as a model system for biopharming

Tomato has been used as a model plant also for production of oral vaccines, which are designed to provide an affordable and easily accessible preventative and curative medical care for the needy. Plant systems-based recombinant proteins and vaccines combine the therapeutic power of antigenic peptides expressed within fruit tissue with low cost of production, safety, easy transportation and availability (Chen et al., 2009; Ahmad et al., 2012; Aryamvally et al., 2016). Some interesting examples of edible tomato vaccines are presented below. Tomato is suitable for production of oral vaccines as fresh edible fruits, because of relatively efficient transformation, stackability of genes via crossing, large-scale greenhouse production and processing technology (cholera toxin B subunit expression, Jani et al., 2002, 2004; Warzecha and

Mason, 2003). Challenges of edible vaccine production in fruits include the presence of suboptimal antigen concentration, as ripe tomato fruits contain <0.7% protein and expression of foreign proteins to high levels is limited (Youm et al., 2008). Use of stronger promoters, tissue-specific promoters, signal peptides and codon optimization have been applied to overcome these challenges (Lauterslager et al., 2001; Sojikul et al., 2003; Tackaberry et al., 2003; Youm et al., 2005). Utilization of strong adjuvants such as β subunit of the cholera toxin, alongside the primary antigen molecule may bolster the immunogenic response (Youm et al., 2008; Baldauf et al., 2015). Nanotechnology is another avenue for novel ways to thwart difficulties in this field in terms of site-specific delivery of the oral vaccine (Zhu and Berzofsky, 2013). Plant products used as a source of pharmaceutical proteins have to be kept separate from mainstream food supply (Warzecha and Mason, 2003).

Tomato was successfully transformed with *E. coli* heat-labile enterotoxin B subunit to produce the LTB (*E. coli* heat-labile enterotoxin B subunit) protein. The protein was found to form active pentamers using an ELISA assay (Loc et al., 2014). Also, transgenic tomatoes engineered to express synthetic DPT (diphtheria, pertussis and tetanus) vaccine as a single gene were successfully used to immunize mice orally (Soria-Guerra et al., 2007). Significant IgA and IgG antibody levels were found in the intestine but response was weaker in tracheopulmonary fluids. The results of this study pointed out the availability of a therapeutic tomato that could replace the traditional triple vaccine in the near future. Transgenic tomato expressing codon-optimized thymosin α I concatamer (an immune system synergist) were tested and found to stimulate the proliferation of mice spleenic lymphocytes. Mature fruits were found to accumulate protein up to 6.1 µg g^{-1} fresh weight (Chen et al., 2009). ORF2 partial gene of hepatitis E virus and large surface antigen gene of hepatitis B virus were engineered in transgenic tomato leaves and fruits as viral antigens (Ma et al., 2003; Lou et al., 2007). A key player in the development of Alzheimer's disease, human β-amyloid protein (Aβ), was overexpressed as trimer to pentamer tandem repeats under the control of CaMV 35S promoter in tomato (Youm et al., 2008). Balb/c mice immunized orally with total soluble extracts from these transgenic tomatoes and boosted by the synthetic Aβ peptide emulsified in alum, elicited an immune response and the immunized mice produced serum antibodies against the Aβ antigen as confirmed by western blots and ELISA. These preliminary results are very promising for developing novel antigens for immunization.

7 Future trends and conclusion

There has been a phenomenal advancement in the biotechnology of agricultural crops. This revolution has greatly modified and reduced the use of pesticides in the production of agronomical crops, particularly corn and soya bean, over 90% of which are now genetically enhanced by biotechnology. As was made apparent here, tomato is a good model system to test various axioms to enhance crop productivity and improve fruit quality, including shelf life and nutritional attributes. These biotech-enhanced crop varieties have gone through laborious field tests for crop performance and production strategies, and yielded huge amounts of data on the use of biotechnology for enhancing crop productivity and producing value-added crops. Thus, biotechnology tools have led to the development of novel tomato genotypes that include enhanced abiotic stress tolerance with a great potential to overcome the present and future challenges imposed by global climate change; improved resistance to biotic stress to reduce devastating losses due to diseases; enhancement in fruit shelf life and quality to reduce post-harvest losses; boosted levels of many phytonutrients with the potential for human health and wellness promotion – examples include folates, anthocyanins, carotenoids and anti-ageing polyamines, particularly to meet recommended daily allowance and reduce physiological disorders in human population; novel plants that can become factories to produce pharmaceuticals including vaccines.

Translation of biotechnology advancements for developing horticultural crops should lead to the emergence of 'super' speciality crops, which we can only imagine today. However, in spite of this extensive research carried out worldwide independently by academicians, public and privately supported researchers, and industry has had limited commercial translation due to unfounded and non-scientific perceptions causing lack of support from retailers and food producers (Aerni, 2013). A recent report published by Committee on Genetically Engineered Crops, The National Academies of Sciences, Engineering and Medicine (2016) analysed about 900 research publications on commercial crops developed through genetic engineering and found positive effects on human health and agriculture (www.nationalacademies.org). No evidence of environmental problems due to cultivation of GE crops was found. Emerging technologies such as CRISPR/Cas9 (Barrangou et al., 2007) and synthetic biology provide great precision for organismal genome improvement, including making single nucleotide changes as do radiation and chemical methods for mutation. The committee noted that both genetic engineering and conventional breeding processes should be evaluated for potential harm to humans and environment. Organized and stringent regulatory system and rigorous risk assessment, for demonstrable safety and efficacy of genetically engineered products and not the process, are central to shifting the attention of the public from the technique per se to the advantages offered by novel traits. It is, however, clear that this technology is here to stay and would greatly help in maintaining food security to ever increasing world population.

8 Where to look for further information

1   Razdan, M. K. and Mattoo, A. K. 2007. *Genetic Improvement of Solanaceous Crops: Volume 2: Tomato*, Science Publishers, Inc. Enfield, USA, p. 451.

2   Nath, N., Bouzayen, M., Mattoo, A. K. and Pech, J.-C. 2014. *Fruit Ripening: Physiology, Signalling and Genomics*, CABI, Oxfordshire, UK, p. 321.

3   Fatima, T., Rivera-Domínguez T.-R., Tiznado-Hernandez, M.-E., Handa, A. K. and Mattoo, A. K. 2008. Tomato. In: Kole, C. and Hall, T. C. (eds.), *Compendium of Transgenic Crop Plants: Transgenic Vegetable Crops*. Wiley-Blackwell Publishing, Oxford, UK, pp. 1–46.

4   Di Matteo, A., Rigano, M. M., Sacco, A., Frusciante, L. and Barone, A. 2011. Genetic transformation in tomato: novel tools to improve fruit quality and pharmaceutical production. In: María Alvarez (ed.), *Genetic Transformation*, InTech, DOI: 10.5772/24521. Available from: http://www.intechopen.com/books/genetic-transformation/genetic-transformation-in-tomato-novel-tools-to-improve-fruit-quality-and-pharmaceutical-production

5   Handa, A. K., Tiznado-Hernandez, M.-E. and Mattoo, A. K. 2012. Fruit development and ripening: a molecular perspective. In: Altman, A. and Hasegawa, P. M. (eds.), *Plant Biotechnology and Agriculture*. Prospects for the 21st Century, Chapter 26, Elsevier, Inc., pp. 405–24.

6   Brooks, C., Nekrasov, V., Lippman, Z. B. and Van Eck, J. 2014. Efficient gene editing in tomato in the first generation using the clustered regularly interspaced short palindromic repeats/CRISPR-associated9 System. *Plant Physiology* 166: 1292–7.

7   Frary, A. and Van Eck, J. 2004. Organogenesis from transformed tomato explants. In: Peña L. (ed.), *Transgenic Plants: Methods and Protocols*, Totowa, NJ: Humana Press, pp. 141–50.

8   Keith, R. 1992. *Safety Assessment of Genetically Engineered Fruits and Vegetables: A Case Study of the Flavr Savr Tomato*. CRC Press.

9 Acknowledgements

A. K. H. research was supported by USDA/NIFA 2010-65115-20374 and USDA/NIFA 2012-67017-30159. Trade names or commercial products mentioned in this publication are only to provide specific information and do not imply any recommendation or endorsement by the authors. USDA is an Equal Employment Opportunity provider.

10 References

Abdeen, A., Virgós, A., Olivella, E., Villanueva, J., Avilés, X., Gabarra, R. and Prat, S. 2005. Multiple insect resistance in transgenic tomato plants over-expressing two families of plant proteinase inhibitors. *Plant Molecular Biology* 57: 189–202.

Abu-El-Heba, G. A., Hussein, G. M. and Abdalla, N. A. 2008. A rapid and efficient tomato regeneration and transformation system. *Landbauforschung Volkenrode* 58: 103–10.

Adato, A., Mandel, T., Mintz-Oron, S., Venger, I., Levy, D., Yativ, M., Dominguez, E., Wang, Z., DeVos, R. C. H., Jetter, R., Schreiber, L., Heredia, A., Rogachev, I. and Aharoni, A. 2009. Fruit surface flavonoid accumulation in tomato is controlled by a SlMYB12-regulated transcriptional network. *PLOS Genetics* 5: e1000777.

Aerni, P. 2013. Resistance to agricultural biotechnology: The importance of distinguishing between weak and strong public attitudes. *Biotechnology Journal* 8: 1129–32.

Agharbaoui, Z., Greer, A. F. and Tabaeizadeh, Z. 1995. Transformation of the wild tomato *Lycopersicon chilense* Dun. by *Agrobacterium tumefaciens*. *Plant Cell Reports* 15: 102–5.

Ahmad, P., Ashraf, M., Younis, M., Hu, X., Kumar, A., Akram, N. A. and Al-Qurainy, F. 2012. Role of transgenic plants in agriculture and biopharming. *Biotechnology Advances* 30: 524–40.

Ainsworth, E. A. and McGrath, J. M. 2010. Direct effects of rising atmospheric carbon dioxide and ozone on crop yields. *Climate Change and Food Security Advances in Global Change Research* 37 (Part II): 109–30. doi:10.1007/978-90- 481-2953-9_7.

Ajenifujah-Solebo, S. O. A., Isu, N. A., Olorode, O., Ingelbrecht, I. and Abiade, O. O. 2012. Tissue culture regeneration of three Nigerian cultivars of tomatoes. *African Journal of Plant Science* 14: 370–5.

Alan, A. R., Blowers, A. and Earle, E. D. 2004. Expression of a magainin-type antimicrobial peptide gene (MSI-99) in tomato enhances resistance to bacterial speck disease. *Plant Cell Reports* 22: 388–96. doi: 10.1007/s00299-003-0702-x.

Alexander, L. and Grierson, D. 2002. Ethylene biosynthesis and action in tomato: A model for climacteric fruit ripening. *Journal of Experimental Botany* 53: 2039–55.

Álvarez-Viveros, M. F., Inostroza-Blancheteau, C., Timmermann, T., Gonzalez, M. and Arce-Johnson, P. 2013. Overexpression of GlyI and GlyII genes in transgenic tomato (*Solanum lycopersicum* Mill.) plants confers salt tolerance by decreasing oxidative stress. *Molecular Biology Reports* 40: 3281–90. doi:10.1007/s11033-012-2403-4.

Amaya, I., Osorio, S., Martinez-Ferri, E., Lima-Silva, V., Doblas, V. G., Fernández-Muñoz, R., Fernie, A. R., Botella, M. A. and Valpuesta, V. 2015. Increased antioxidant capacity in tomato by ectopic expression of the strawberry D-galacturonate reductase gene. *Biotechnology Journal* 10: 490–500. doi:10.1002/biot.201400279.

Anwar, R., Mattoo, A. K. and Handa, A. K. 2015. Polyamine interactions with plant hormones: Crosstalk at several levels. In: Kusano, T. and Suzuki, H. (eds.), *Polyamines: A Universal Molecular Nexus for Growth, Survival, and Specialized Metabolism*. Tokyo: Springer, Japan, pp. 267–302.

Arora, A., Sairam, R. K. and Srivastava, G. C. 2002. Oxidative stress and antioxidative system in plants. *Current Science* 82: 1227–38.

Arrillaga, I., Gil-Mascarell, R., Gisbert, C., Sales, E., Montesinos, C., Serrano, R. and Moreno, V. 1998. Expression of the yeast HAL2 gene in tomato increases the *in vitro* salt tolerance of transgenic progenies. *Plant Science* 136: 219–26. doi:10.1016/S0168-9452(98)00122-8.

Aryamvally, A., Gunasekaran, V., Narenthiran K. R. and Pasupathi, R. 2016. New strategies toward edible vaccines: An overview. *Journal of Dietary Supplements* 11: 1–16.

Bai, Y. and Lindhout, P. 2007. Domestication and breeding of tomatoes: What have we gained and what can we gain in the future? *Annals of Botany* 100: 1085–94.

Baldauf, K. J., Royal, J. M., Hamorsky, K. T. and Matoba, N. 2015. Cholera Toxin B: One subunit with many pharmaceutical applications. *Toxins* 7: 974–96.

Barrangou, R., Fremaux, C., Deveau, H., Richards, M., Boyaval, P., Moineau, S., Romero, D. A. and Horvath, P. 2007. CRISPR provides acquired resistance against viruses in prokaryotes. *Science* 315: 1709–12.

Barthakur, S., Babu, V. and Bansal, K. C. 2001. Over-expression of osmotin induces proline accumulation and confers tolerance to osmotic stress in transgenic tobacco. *Journal of Plant Biochemistry and Biotechnology* 10: 31–7.

Bartoszewski, G., Niedziela, A., Szwacka, M. and Niemirowicz-Szczytt, K. 2003. Modification of tomato taste in transgenic plants carrying a thaumatin gene from *Thaumatococcus danielli Benth*. *Plant Breeding* 122: 347–51.

Bergougnoux, V. 2014. The history of tomato: From domestication to biopharming. *Biotechnology Advances* 32: 170–89.

Bevan, M. W., Flavell, R. B. and Chilton, M.-D. 1983. A chimaeric antibiotic resistance gene as a selectable marker for plant cell transformation. *Nature* 304: 184–7.

Bhaskaran, S. and Savithramma, D. L. 2011. Co-expression of *Pennisetum glaucum* vacuolar Na$^+$/H$^+$ antiporter and *Arabidopsis* H$^+$-pyrophosphatase enhances salt tolerance in transgenic tomato. *Journal of Experimental Botany* 62: 5561–70. doi:10.1093/jxb/err237.

Bohner, J. and Bangerth, F. 1988. Cell number, cell size and hormone levels in semi-isogenic mutants of *Lycopersicon pimpinellifolium* differing in fruit size. *Physiologia Plantarum* 72: 316–20.

Bovy, A., de Vos, R., Kemper, M., Schijlen, E., Almenar Pertejo, M., Muir, S., Collins, G., Robinson, S., Verhoeyen, M., Hughes, S., Santos-Buelga, C. and van Tunen, A. 2002. High-flavonol tomatoes resulting from the heterologous expression of the maize transcription factor genes LC and C1. *Plant Cell Online* 14: 2509–26.

Bovy, A., Schijlen, E. and Hall, R. D. 2007. Metabolic engineering of flavonoids in tomato (*Solanum lycopersicum*): The potential for metabolomics. *Metabolomics* 3: 399–412.

Butelli, E., Titta, L., Giorgio, M., Mock, H.-P., Matros, A., Peterek, S., Schijlen EGWM., Hall, R. D., Bovy, A. G., Luo, J. and Martin, C. 2008. Enrichment of tomato fruit with health-promoting anthocyanins by expression of select transcription factors. *Nature Biotechnology* 26: 1301–8.

Catanzariti A.-M., Lim, G. T. T. and Jones, D. A. 2015. The tomato I-3 gene: A novel gene for resistance to *Fusarium* wilt disease. *New Phytologist* 207: 106–18.

Cebolla, A., Maria Vinardell, J., Kiss, E., Olah, B., Roudier, F., Kondorosi, A. and Kondorosi, E. 1999. The mitotic inhibitor ccs52 is required for endoreduplication and ploidy-dependent cell enlargement in plants. EMBO Journal 18: 4476–84.

Chan, Y. L., He, Y., Hsiao, T. T., Wang, C. J., Tian, Z. and Yeh, K. W. 2015. Pyramiding taro cystatin and fungal chitinase genes driven by a synthetic promoter enhances resistance in tomato to root-knot nematode Meloidogyne incognita. Plant Science 231: 74–81.

Chaudry, A., Abbas, S., Yasmin, A., Rashid, H., Ahmed, H. and Anjum, M. A. 2010. Tissue culture studies in tomato (Lycopersicon esculentum) var. Moneymaker. Pakistan Journal of Botany 1: 155–63.

Chen, G., Hackett, R., Walker, D., Taylor, A., Lin, Z., and Grierson, D. 2004. Identification of a specific isoform of tomato lipoxygenase (TomloxC) involved in the generation of fatty acid derived flavor compounds. Plant Physiology 136: 2641–51.

Chen, H., Wilkerson, C. G., Kuchar, J. A., Phinney, B. S. and Howe, G. A. 2005. Jasmonate-inducible plant enzymes degrade essential amino acids in the herbivore midgut. Proceedings of the National Academy of Sciences USA 102: 19237–42.

Chen, Y., Wang, A., Zhao, L., Shen, G., Cui, L. and Tang, K. 2009. Expression of thymosin α1 concatemer in transgenic tomato (Solanum lycopersicum) fruits. Biotechnology and Applied Biochemistry 52: 303–12.

Cheng, L., Zou, Y., Ding, S., Zhang, J., Yu, X., Cao, J. and Lu, G. 2009. Polyamine accumulation in transgenic tomato enhances the tolerance to high temperature stress. Journal of Integrative Plant Biology 51: 489–99.

Cheniclet, C., Rong, W. Y., Causse, M., Frangne, N., Bolling, L., Carde, J.-P. and Renaudin, J.-P. 2005. Cell expansion and endoreduplication show a large genetic variability in pericarp and contribute strongly to tomato fruit growth. Plant Physiology 139: 1984–94.

Chinnusamy, V., Ohta, M., Kanrar, S., Lee, B., Hong, X., Agarwal, M. and Zhu, J. K. 2003. ICE1: A regulator of cold-induced transcriptome and freezing tolerance in Arabidopsis. Genes and Development 17:1043–54.

Choi, J. Y., Seo, Y. S., Kim, S. J., Kim, W. T. and Shin, J. S. 2011. Constitutive expression of CaXTH3, a hot pepper xyloglucan endotransglucosylase/hydrolase, enhanced tolerance to salt and drought stresses without phenotypic defects in tomato plants (Solanum lycopersicum cv. Dotaerang). Plant Cell Reports 30(5): 867–77. doi:10.1007/s00299-010-0989-3.

Chyi, Y. S. and Phillips, G. C. 1987. High efficiency Agrobacterium mediated transformation of Lycopersicon based on conditions favorable for regeneration. Plant Cell Reports 6: 105–8.

Coego, A., Ramirez, V., Ellul, P., Mayda, E. and Vera, P. 2005. The H$_2$O$_2$-regulated Ep5C gene encodes a peroxidase required for bacterial speck susceptibility in tomato. Plant Journal 42: 283–93.

Colliver, S., Bovy, A., Collins, G., Muir, S., Robinson, S., de Vos CHR. and Verhoeyen, M. E. 2002. Improving the nutritional content of tomatoes through reprogramming their flavonoid biosynthetic pathway. Phytochemistry Reviews 1: 113–23.

Cortina, C. and Culianez-Macia, F. A. 2004. Tomato transformation and transgenic plant production. Plant Cell, Tissue and Organ Culture 76: 269–75.

Cortina, C. and Culianez-Macia, F. A. 2005. Tomato abiotic stress enhanced tolerance by trehalose biosynthesis. Plant Science 169: 75–82.

Crowe, J. H., Crowe, L. M. and Chapman, D. 1984. Preservation of membranes in anhydrobiotic organisms: The role of trehalose. Science 223: 701–3.

D'Ambrosio, C., Giorio, G., Marino, I., Merendino, A., Petrozza, A., Salfi, L., Stigliani, A. L. and Cellini, F. 2004. Virtually complete conversion of lycopene into E-carotene in fruits of tomato plants transformed with the tomato lycopene β-cyclase (tlcy-b) cDNA. Plant Science 166: 207–14.

D'Introno, A., Paradiso, A., Scoditti, E., D'Amico, L., De Paolis, A., Carluccio, M. A., Nicoletti, I., DeGara, L., Santino, A. and Giovinazzo, G. 2009. Antioxidant and anti-infl ammatory properties of tomato fruits synthesizing different amounts of stilbenes. Plant Biotechnology Journal 7: 422–9.

Dal Cin, V., Tieman, D. M., Tohge, T., McQuinn, R., de Vos, R. C. H., Osorio, S., Schmelz, E. A., Taylor, M. G., Smits-Kroon, M. T., Schuurink, R. C., Haring, M. A., Giovannoni, J., Fernie, A. R. and Klee, H. J. 2011. Identification of genes in the phenylalanine metabolic pathway by ectopic expression of a MYB transcription factor in tomato fruit. Plant Cell Online 23: 2738–53.

Davidovich-Rikanati, R., Lewinsohn, E., Bar, E., Iijima, Y., Pichersky, E. and Sitrit, Y. 2008. Overexpression of the lemon basil alpha-zingiberene synthase gene increases both mono- and sesquiterpene contents in tomato fruit. *The Plant Journal :For Cell and Molecular Biology* 56: 228–38. doi:10.1111/j.1365-313X.2008.03599.x.

Davidovich-Rikanati, R., Sitrit, Y., Tadmor, Y., Iijima, Y., Bilenko, N., Bar, E., Carmona, B., Fallik, E., Dudai, N., Simon, J. E., Pichersky, E. and Lewinsohn, E. 2007. Enrichment of tomato flavor by diversion of the early plastidial terpenoid pathway. *Nature Biotechnology* 25: 899–901.

Davis, M. E., Lineberger, R. D. and Miller, A. R. 1991. Effects of tomato cultivar, leaf age, and bacterial strain on transformation by *Agrobacterium tumefaciens*. *Plant Cell, Tissue and Organ Culture* 24:115–21.

Davuluri, G. R., van Tuinen, A., Fraser, P. D., Manfredonia, A., Newman, R., Burgess, D., Brummell, D. A., King, S. R., Palys, J., Uhlig, J., Bramley, P. M., Pennings, H. M. J. and Bowler, C. 2005. Fruitspecific RNAi-mediated suppression of *DET1* enhances carotenoid and flavonoid content in tomatoes. *Nature Biotechnology* 23: 890–5.

De Giovanni, C., Dell'Orco, P., Bruno, A., Ciccarese, F., Lotti, C. and Ricciardi, L. 2004. Identification of PCR-based markers (RAPD, AFLPB) linked to a novel powdery mildew resistance gene (ol-2) in tomato. *Plant Science* 166: 41–8.

Delannay, X., LaVallee, B. J., Proksch, R. K., et al. 1989. Field performance of transgenic tomato plants expressing the *Bacillus thuringiensis* var. kurstaki insect control protein. *Bio/Technology* 7: 1265–9.

Dharmapuri, S., Rosati, C., Pallara, P., Aquilani, R., Bouvier, F., Camara, B. and Giuliano, G. 2002. Metabolic engineering of xanthophyll content in tomato fruits. *FEBS Letters* 519: 30–4.

Domínguez, T., Hernández, M. L., Pennycooke, J. C., Jiménez, P., Martínez-Rivas, J. M., Sanz, C., Stockinger, E. J., Sánchez-Serrano, J. J. and Sanmartín, M. 2010. Increasing ω-3 desaturase expression in tomato results in altered aroma profile and enhanced resistance to cold stress. *Plant Physiology* 153: 655–65.

Economic Research Service, United States Department of Agriculture. 2016. http://www.ers.usda.gov/topics/crops/vegetables-pulses/tomatoes.aspx.

Elad, Y., Rav-David, D., Leibman, D., Vintal, H., Vunsh, R., Moorthy, H., Gal-On, A. and Loebenstein, G. 2012. Tomato plants transformed with the inhibitor-of-virus-replication gene are partially resistant to several pathogenic fungi. *Annals of Applied Biology* 161: 16–23.

Ellul, P., Garcia-Sogo, B., Pineda, B., Rios, G., Roig, L. A. and Moreno, V. 2003. The ploidy level of transgenic plants in *Agrobacterium*-mediated transformation of tomato cotyledons (*Lycopersicon esculentum* Mill.) is genotype and procedure dependent. *Theoretical and Applied Genetics* 106: 231–8.

El-Siddig, M. A., El-Hussein, A. A. and Saker, M. M. 2011. *Agrobacterium*-mediated transformation of tomato plants expressing defensin gene. *International Journal of Agriculture Research* 6: 323–34.

Enfissi, E. M. A., Fraser, P. D., Lois, L.-M., Boronat, A., Schuch, W. and Bramley, P. M. 2005. Metabolic engineering of the mevalonate and non-mevalonate isopentenyl diphosphate-forming pathways for the production of health-promoting isoprenoids in tomato. *Plant Biotechnology Journal* 3: 17–27.

FAOSTAT. 2015. Food and Agriculture Organization of the United Nations, Statistics Division. http://faostat3.fao.org.

Fatima T., Handa, A. K. and Mattoo, A. K. 2013. Functional foods: Genetics, metabolome, and engineering phytonutrient levels. In: Ramawat, K. G. and Mérillon, J. M. (eds.), *Natural Products*. doi 10.1007/978-3-642-22144-6_50, Springer-Verlag Berlin Heidelberg, pp. 1715–49.

Fatima, T., Rivera-Domínguez M., Troncoso-Rojas R., Tiznado-Hernández M. E., Handa, A. K. and Mattoo, A. K. 2008. Tomato. In: Kole, C. and Hall, T. C. (eds.), *Compendium of Transgenic Crop Plants: Transgenic Vegetable Crops*. Wiley-Blackwell Publishing, Oxford, UK, pp. 1–46.

Fiocchetti, F., Caruso, C., Bertini, L., Vitti, D., Saccardo, F. and Tucci, M. 2006. Over-expression of a pathogenesis-related protein gene in transgenic tomato alters the transcription patterns of other defence genes. *Journal of Horticultural Science and Biotechnology* 81: 27–32.

Fluhr, R. and Mattoo, A. K. 1996. Ethylene - biosynthesis and perception. *Critical Reviews in Plant Sciences* 15: 479–523. doi 10.1080/07352689609382368.

Foolad, M. R. 2004. Recent advances in genetics of salt tolerance in tomato. *Plant Cell, Tissue and Organ Culture* 76: 101–19.

Forkmann, G. and Heller, W. 1999. Biosynthesis of flavonoids. In: Barton, D., Nakanishi, K. and Meth-Cohn, O. (eds.), *Comprehensive natural products chemistry.* Elsevier, Amsterdam, pp. 713–48.

Fraley, R. T., Rogers, S. G., Horsch, R. B., et al. 1983. Expression of bacterial genes in plant cells. *Proceedings of the National Academy of Sciences USA* 80: 4803–7.

Fraser, P. D., Enfissi, E. M. A., Halket, J. M., Truesdale, M. R., Yu, D., Gerrish, C. and Bramley, P. M. 2007. Manipulation of phytoene levels in tomato fruit: Effects on isoprenoids, plastids, and intermediary metabolism. *Plant Cell Online* 19: 3194–211.

Fraser, P. D., Römer, S., Shipton, C. A., Mills, P. B., Kiano, J. W., Misawa, N., Drake, R. G., Schuch, W. and Bramley, P. M. 2002. Evaluation of transgenic tomato plants expressing an additional phytoene synthase in a fruit-specific manner. *Proceedings of the National Academy of Sciences USA* 99: 1092–7.

Fray, R. G., Wallace, A., Fraser, P. D., Valero, D., Hedden, P., Bramley, P. M. and Grierson, D. 1995. Constitutive expression of a fruit phytoene synthase gene in transgenic tomatoes causes dwarfism by redirecting metabolites from the gibberellin pathway. *Plant Journal* 8: 693–701.

Fuentes, A., Ramos, P. L., Elvira Fiallo, E., Callard, D., Sánchez, Y., Peral, R., Rodríguez, R. and Pujol, M. 2006. Intron-hairpin RNA derived from replication associated protein C1 gene confers immunity to tomato yellow leaf curl virus infection in transgenic tomato plants. *Transgenic Research* 15: 291–304.

Gal, T. Z., Aussenberg, E. R., Burdman, S., Kapulnik, Y. and Koltai, H. 2006. Expression of a plant expansin is involved in the establishment of root knot nematode parasitism in tomato. *Planta* 224: 155–62.

Giliberto, L., Perrotta, G., Pallara, P., Weller, J. L., Fraser, P. D., Bramley, P. M., Fiore, A., Tavazza, M. and Giuliano, G. 2005. Manipulation of the blue light photoreceptor Cryptochrome 2 in tomato affects vegetative development, flowering time, and fruit antioxidant content. *Plant Physiology* 137: 199–208.

Giovannoni, J. J. 2004. Genetic regulation of fruit development and ripening. *Plant Cell* 16: S170–S180.

Giovannoni, J. J. 2007. Fruit ripening mutants yield insights into ripening control. *Current Opinion in Plant Biology* 10: 283–9.

Giovinazzo, G., D'Amico, L., Paradiso, A., Bollini, R., Sparvoli, F. and DeGara, L. 2005. Antioxidant metabolite profiles in tomato fruit constitutively expressing the grapevine stilbene synthase gene. Plant *Biotechnology Journal* 3: 57–69.

Gisbert, C., Rus, A. M., Bolarín, M. C., López-Coronado, J. M., Arrillaga, I., Montesinos, C., Caro, M., Serrano, R. and Moreno, V. 2000. The yeast HAL1 gene improves salt tolerance of transgenic tomato. *Plant Physiology* 123(1): 393–402.

Goel, D., Singh, A. K., Yadav, V., Babbar, S. B. and Bansal, K. C. 2010. Overexpression of osmotin gene confers tolerance to salt and drought stresses in transgenic tomato (*Solanum lycopersicum* L.). *Protoplasma* 245: 133–41.

Goel, D., Singh, A. K., Yadav, V., Babbar, S. B., Murata, N. and Bansal, K. C. 2011. Transformation of tomato with a bacterial codA gene enhances tolerance to salt and water stresses. *Journal of Plant Physiology* 168: 1286–94.

Goldsbrough, A. P., Tong, Y. and Yoder, J. I. 1996. Lc as a non-destructive visual reporter and transposition excision marker gene for tomato. *Plant Journal* 9: 927–33.

Gong, Z.-Z., Yamagishi, E., Yamazaki, M. and Saito, K. 1999. A constitutively expressed Myc-like gene involved in anthocyanin biosynthesis from *Perilla frutescens*: Molecular characterization, heterologous expression in transgenic plants and transactivation in yeast cells. *Plant Molecular Biology* 41: 33–44.

Gonzalez, N., Gevaudant, F., Hernould, M., Chevalier, C. and Mouras, A. 2007. The cell cycle associated protein kinase WEE1 regulates cell size in relation to endoreduplication in developing tomato fruit. *Plant Journal* 51: 642–55.

Good, X., Kellogg, J. A., Wagonner, W., Langhoff, D., Matsumura, W. and Bestwick, R. K. 1994. Reduced ethylene synthesis by transgenic tomato expressing S-adenosylmethionine hydrolase. *Plant Molecular Biology* 26: 781–90.

Goyal, R. K., Fatima, T., Topuz, M., Bernadec, A., Sicher, R., Handa, A. K. and Mattoo, A. K. 2016. Pathogenesis-related protein 1b1 (PR1b1) is a major tomato fruit protein responsive to chilling temperature and upregulated in high polyamine transgenic genotypes. *Frontiers in Plant Science*, in press.

Grant, J. J. and Loake, G. J. 2000. Role of reactive oxygen intermediates and cognate redox signaling in disease resistance. *Plant Physiology* 124: 21–9.

Guo, F., Zhou, W., Zhang, J., Xu, Q. and Deng, X. 2012. Effect of the citrus lycopene β-cyclase transgene on carotenoid metabolism in transgenic tomato fruits. *PLOS One* 7: e32221.

Hamilton, A. J., Lycett, G. W. and Grierson, D. 1990. Antisense gene that inhibits synthesis of the hormone ethylene in transgenic plants. *Nature* 6281: 284–7.

Han, H. Q., Liu, Y., Jiang, M. M., Ge, H. Y. and Chen, H. Y. 2015. Identification and expression analysis of YABBY family genes associated with fruit shape in tomato (*Solanum lycopersicum* L.). *Genetics and Molecular Research* 14: 7079–91.

Handa A. K. and Mattoo, A. K. 2010. Differential and functional interactions emphasize the multiple roles of polyamines in plants. *Plant Physiology and Biochemistry* 48: 540–6.

Handa, A. K., Srivastava, A., Deng, Z., Gaffe, J., Arora, A., Tiznado-Hernández, M.-E., Goyal, R. K., Malladi, A., Negi, P. S. and Mattoo, A. K. 2010. Biotechnological interventions to improve plant developmental traits. In: Kole, C., Michler, C. H., Abbott, A. G. and Hall, T. C. (eds.), *Transgenic Crop Plants*. Springer, Berlin Heidelberg, pp. 199–248.

Handa, A. K., Tiznado-Hernández, M.-E. and Mattoo, A. K. 2012. Fruit development and ripening: A molecular perspective. In: Altman, A. and Hasegawa, P. M. (eds.), *Plant Biotechnology and Agriculture: Prospects for 21st Century*. Elsevier, New York, NY, pp. 405–24.

Hanur, V. S., Reddy, B., Arya, V. V. and Rami Reddy, P. V. 2015. Genetic transformation of tomato using Bt Cry2A gene and characterization in Indian cultivar Arka Vikas. *Journal of Agricultural Science and Technology* 17: 1805–14.

Hasan, M., Khan, A. J., Khan, S., Shah, A. H., Khan, A. R. and Mirza, B. 2008. Transformation of tomato (*Lycopersicon esculentum* Mill.) with *Arabidopsis* early flowering gene *APETALI (API)* through *Agrobacterium* infiltration of ripened fruits. *Pakistan Journal of Botany* 1: 161 73.

Herbette, S., Le Menn, A., Rousselle, P., Ameglio, T., Faltin, Z., Branlard, G., Eshdat, Y., Julien, J. L., Drevet, J. R. and Roeckel-Drevet, P. 2005. Modification of photosynthetic regulation in tomato overexpressing glutathione peroxidase. *Biochimica et Biophysica Acta* 1724: 108–18.

Hererra-Estrella, L., Depicker, A., Montagu, M. V. and Schell. 1983. Expression of chimaeric genes transferred into plant cells using a Ti-plasmid-derived vector. *Nature* 303: 209–13.

Himabindu, Y., Chakradhar, T., Reddy, M. C., Kanygin, A., Redding, K. E. and Chandrasekhar, T. 2016. Salt-tolerant genes from halophytes are potential key players of salt tolerance in glycophytes. *Environmental and Experimental Botany* 124: 39–63.

Hseih, T. H., Lee, J. T., Charng, Y. Y. and Chan, M. T. 2002. Tomato plants ectopically expressing *Arabidopsis CBF1* show enhanced resistance to water deficit stress. *Plant Physiology* 130: 618–26.

Hu D-G., Ma Q-J., Sun C-H., Sun M.-H., You C-X. and Hao Y.-J. 2016. Overexpression of MdSOS2L1, a CIPK protein kinase, increases the antioxidant metabolites to enhance salt tolerance in apple and tomato. *Physiologia Plantarum* 156: 201–14.

Huang X-S., Zhang, Q., Zhu, D., Fu, X., Wang, M., Zhang, Q., Moriguchi, T. and Liu, J.-H. 2015. ICE1 of *Poncirus trifoliata* functions in cold tolerance by modulating polyamine levels through interacting with arginine decarboxylase. *Journal of Experimental Botany*. doi: 10.1093/jxb/erv138.

Huertas, R., Olías, R., Eljakaoui, Z., Gálvez, F. J., Li, J., Alvarez de Morales, P., Belver, A. and Rodríguez-Rosales, M. P. 2012. Overexpression of SlSOS2 (SlCIPK24) confers salt tolerance to transgenic tomato. *Plant, Cell and Environment* 35: 1467–82.

Huertas, R., Rubio, L., Cagnac, O., et al. 2013. The K⁺/H⁺ antiporter LeNHX2 increases salt tolerance by improving K⁺ homeostasis in transgenic tomato. *Plant, Cell and Environment* 36: 2135–49. doi:10.1111/pce.12109.

Hunt, G. M. and Baker, E. A. 1980. Phenolic constituents of tomato fruit cuticles. *Phytochemistry* 19: 1415–19.

Itkin, M., Seybold, H., Breitel, D., Rogachev, I., Meir, S. and Aharoni, A. 2009. TOMATO AGAMOUS-LIKE 1 is a component of the fruit ripening regulatory network. *Plant Journal* 60: 1081–95.

Jabs, T. 1999. Reactive oxygen intermediates as mediators of programmed cell death in plants and animals. *Biochemical Pharmacology* 57: 231–45.

James, C. 2015. Global Status of Commercialized Biotech/GM Crops: 2015. ISAAA Brief No. 51. ISAAA: Ithaca, NY.

Jani, D., Meena, L. S., Rizwan-Ul-Haq, Q. M., Singh, Y., Sharma, A. K. and Tyagi, A. K. 2002: Expression of cholera toxin B subunit in transgenic tomato plants. *Transgenic Research* 11: 47–54.

Jani, D., Singh, N. K., Bhattacharya, S., Meena, L. S., Singh, Y., Upadhyay, S. N., Sharma, A. K. and Tyagi, A. K. 2004. Studies on the immunogenic potential of plant-expressed cholera toxin B subunit. *Plant Cell Reports* 22: 471–7.

Jia, G.-X., Zhu, Z.-Q., Chang, F.-Q. and Li, Y.-X. 2002. Transformation of tomato with the *BADH* gene from *Atriplex* improves salt tolerance. *Plant Cell Reports* 21(2): 141–6. doi:10.1007/s00299–002-0489-1.

Jongedijk, E., Tigelaar, H., Van Roekel J. S. C., Bres-Vloemans, S. A., Dekker, I., Van den Elzen, P. J. M., Cornelissen, B. J. C. and Melchers, L. S. 1995. Synergistic activity of chitinases and β-1,3-glucanases enhances fungal resistance in transgenic tomato plants. *Euphytica* 85: 173–80.

Juan, J. X., Yu, X. H., Jiang, X. M., et al. 2015. *Agrobacterium*-mediated transformation of tomato with the ICE1 transcription factor gene. *Genetics and Molecular Research* 14: 597–608.

Jung Y-J. 2013. Enhanced resistance to bacterial pathogen in transgenic tomato plants expressing cathelicidin antimicrobial peptide. *Biotechnology and Bioprocess Engineering* 18: 615–24. doi: 10.1007/s12257-013-0392-3.

Kamthan, A., Kamthan, M., Azam, M., Chakraborty, N., Chakraborty, S. and Datta, A. 2012. Expression of a fungal sterol desaturase improves tomato drought tolerance, pathogen resistance and nutritional quality. *Scientific Reports* 2: 951.

Kaur, B., Purkayastha, S., Dilbaghi, N. and Chaudhury, A. 2005. Characterization of *Xanthomonas axonopodis* pv. *cyamopsidis*, the bacterial blight pathogen of cluster bean, using PCR-based molecular markers. *Journal of Phytopathology* 153: 470–9.

Kaur, P. and Bansal, K. C. 2010. Efficient production of transgenic tomatoes via *Agrobacterium*-mediated transformation. *Biologia Plantarum* 54: 344–8.

Kausch, K. D., Sobolev, A. P., Goyal, R. K., Fatima, T., Laila-Beevi, R., Saftner, R. A., Handa, A. K. and Mattoo, A. K. 2012. Methyl jasmonate deficiency alters cellular metabolome, including the aminome of tomato (*Solanum lycopersicum* L.) fruit. *Amino Acids* 42: 843–56.

Khaliluev, M. R. and Shpakovskii, G. V. 2013. Genetic engineering strategies for enhancing tomato resistance to fungal and bacterial pathogens. *Russian Journal of Plant Physiology* 60: 721–32. doi: 10.1134/S1021443713050087.

Khare, N., Goyary, D., Singh, N. K., Shah, P., Rathore, M., Anandhan, S., Sharma, D., Arif, M. and Ahmed, Z. 2010. Transgenic tomato cv. Pusa Uphar expressing a bacterial mannitol-1-phosphate dehydrogenase gene confers abiotic stress tolerance. *Plant Cell, Tissue and Organ Culture* 103: 267–77.

Klee, H. J. and Giovannoni, J. J. 2011. Genetics and control of tomato fruit ripening and quality attributes. *Annual Review of Genetics* 45: 41–59.

Klee, H. J., Hayford, M. B., Kretzmer, K. A., Barry, G. F. and Kishore, G. M. 1991. Control of ethylene synthesis by expression of a bacterial enzyme in transgenic tomato plants. *Plant Cell* 3: 1187–93.

Knight, M. R. and Knight, H. 2012. Low-temperature perception leading to gene expression and cold tolerance in higher plants. *New Phytologist* 195: 737.

Kramer, M., Sanders, R., Bolkan, H., Waters, C., Sheehy, R. and Hiatt, W. 1992. Post-harvest evaluation of transgenic tomatoes with reduced levels of polygalacturonase: Processing, firmness and disease resistance. *Post Harvest Biology and Technology* 1: 241–55.

Kramer, M., Sanders, R., Sheehy, R., Melis, M., Kuehn, M. and Hiatt, W. 1990. Field evaluation of tomatoes with reduced polygacturonase by antisense RNA. In: Bennett, A. and O'Neill, S. (eds.), *Horticultural Biotechnology*. New York, Wiley-Liss, Inc., pp. 347–55.

Kramer, M. G. and Redenbaugh, K. 1994. Commercialization of a tomato with an antisense polygalacturonase gene: The FLAVR SAVR™ tomato story. *Euphytica* 79: 293–7.

Kramer, P. and Boyer, J. 1995. *Water Relations of Plants and Soils*. San Diego: Academic Press.

Krieger, E. K., Allen, E., Gilbertson, L. A., Roberts, J. K., Hiatt, W. and Sanders, R. A. 2008. The Flavr Savr tomato, an early example of RNAi technology. *HortScience* 43: 962–4.

Ku, H. M., Doganlar, S., Chen, K. Y. and Tanksley, S. D. 1999. The genetic basis of pear-shaped tomato fruit. *Theoretical and Applied Genetics* 99: 844–50.

Kumar, S. V., Sharma, M. L. and Rajam, M. V. 2006. Polyamine biosynthetic pathway as a novel target for potential applications in plant biotechnology. *Physiology and Molecular Biology of Plants* 12: 13–28.

Kumar, V., Abdul-Baki, A., Anderson, J. D. and Mattoo, A. K. 2005. Cover crop residues enhance growth, improve yield and delay leaf senescence in greenhouse-grown tomatoes. *HortScience* 40: 1307–11.

Kumar, V., Mills, D. J., Anderson, J. D. and Mattoo, A. K. 2004. An alternative agriculture system is defined by a distinct expression profile of select gene transcripts and proteins. *Proceedings of the National Academy of Sciences USA* 101: 10535–40.

Lauterslager TGM., Florack DEA., van der Wal, T. J., Molthoff, J. W., Langeveld JPM., Bosch, D., Boersma WJA., Hilgers LAT. 2001. Oral immunisation of naive and primed animals with transgenic potato tubers expressing LT-B. *Vaccine* 19: 2749–55.

Lawrence, S. D. and Novak, N. G. 2006. Expression of poplar chitinase in tomato leads to inhibition of development in colorado potato beetle. *Biotechnology Letters* 28: 593–9.

Lee, H. I. and Raikhel, N. V. 1995. Prohevein is poorly processed but shows enhanced resistance to a chitin-binding fungus in transgenic plants. *Brazilian Journal of Medical and Biological Research* 28: 743–50.

Lee J.-T., Prasad, V., Yang P.-T.,Wu, J. F., Ho THD., CHarng, Y. Y. and Chan, M. T. 2003a. Expression of *Arabidopsis* CBF1 regulated by an ABA/stress inducible promoter in transgenic tomato confers stress tolerance without affecting yield. *Plant, Cell and Environment* 26: 1181–90. doi: 10.1046/j.1365-3040.2003.01048.x.

Lee, O. S., Lee, B., Park, N., Koo, J. C., Kim, Y. H., Prasad, D. T., Karigar, C., Chun, H. J., Jeong, B. R., Kim, D. H., Nam, J., Yun, J. G., Kwak, S. S., Cho, M. J. and Yun, D. J. 2003b. PnAMPs, the hevein like proteins from *Pharbitis nil* confers disease resistance against phytopathogenic fungi in tomato, *Lycopersicum esculentum* Phytochemistry 62: 1073–9.

Lee, T. J., Coyne, D. P., Clemente, T. E. and Mitra, A. 2002. Partial resistance to bacterial wilt in transgenic tomato plants expressing antibacterial Lactoferrin gene. *Journal of the American Society of Horticultural Science* 127:158–64.

Lewinsohn, E., Schalechet, F., Wilkinson, J., Matsui, K., Tadmor, Y., Nam, K.-H., Amar, O., Lastochkin, E., Larkov, O., Ravid, U., Hiatt, W., Gepstein, S. and Pichersky, E. 2001. Enhanced levels of the aroma and fl avor compound S-linalool by metabolic engineering of the terpenoid pathway in tomato fruits. *Plant Physiology* 127: 1256–65.

Li, C., Feng, X. X., Zhang, Z. P., Sun, X. E. and Wang, L. J. 2012. Studies on salt tolerance in tomato plants by transformation of yHem1. *Acta Horticulturae Sinica Issue* 10: 1937–48.

Li, D., Jiang, X. and Yu, X. 2010. CBF gene cloning and expression analysis. *Plant Physiology Communications* 46: 245–8.

Li, L. and Steffens, J. C. 2002. Overexpression of polyphenol oxidase in transgenic tomato plants results in enhanced bacterial disease resistance. *Planta* 215: 239–47.

Lim, M. Y., Jeong, B. R., Jung, M. and Harn, C. H. 2016. Transgenic tomato plants expressing strawberry d-galacturonic acid reductase gene display enhanced tolerance to abiotic stresses. *Plant Biotechnology Reports* 10: 105–16.

Lin, D., Wei, Y. and Wang, S. 2000. Tomato resistance to low temperature research progress. *Journal of Shenyang Agricultural University* Issue 6: 585–9.

Lin, W. C., Lu, C. F., Wu, J. W., Cheng, M. L., Lin, Y. M., Yang, N. S., et al. 2004. Transgenic tomato plants expressing the *Arabidopsis NPR1* gene display enhanced resistance to a spectrum of fungal and bacterial diseases. *Transgenic Research* 13: 567–81.

Lippman, Z. and Tanksley, S. D. 2001. Dissecting the genetic pathway to extreme fruit size in tomato using a cross between the small-fruited wild species *Lycopersicon pimpinellifolium* and *L. esculentum* var. Giant Heirloom. *Genetics* 158: 413–22.

Lison, P., Rodrigo, I. and Conejero, V. 2006. A novel function for the cathepsin D inhibitor in tomato. *Plant Physiology* 142: 1329–39.

Liu, J., Van Eck, J., Cong, B. and Tanksley, S. D. 2002. A new class of regulatory genes underlying the cause of pear-shaped tomato fruit. *Proceedings of the National Academy of Sciences USA* 99: 13302–6.

Liu, J. H., Kitashiba, H., Wang, J., Ban, Y. and Moriguchi, T. 2007. Polyamines and their ability to provide environmental stress tolerance to plants. *Plant Biotechnology* 24: 117–26.

Liu, Y., Roof, S., Ye, Z., Barry, C., van Tuinen, A., Vrebalov, J., Bowler, C. and Giovannoni, J. 2004. Manipulation of light signal transduction as a means of modifying fruit nutritional quality in tomato. *Proceedings of the National Academy of Sciences USA* 101: 9897–902.

Loc, N. H., Long, D. T., Kim, T. G. and Yang, M. S. 2014. Expression of *Escherichia coli* heat-labile enterotoxin B subunit in transgenic tomato (*Solanum lycopersicum* L.) fruit. *Czech Journal of Genetics and Plant Breeding* 50: 26–31.

Lou, X. M., Yao, Q. H., Zhang, Z., Peng, R. H., Xiong, A. S. and Wang, H. K. 2007. Expression of the human hepatitis B virus large surface antigen gene in transgenic tomato plants. *Clinical and Vaccine Immunology* 14: 464–9.

Lu, Y., Rijzaani, H., Karcher, D., Ruf, S. and Bock, R. 2013. Efficient metabolic pathway engineering in transgenic tobacco and tomato plastids with synthetic multigene operons. *Proceedings of the National Academy of Sciences of the United States of America* 110(8): E623–32. doi:10.1073/pnas.1216898110.

Luo, J., Butelli, E., Hill, L., Parr, A., Niggeweg, R., Bailey, P., Weisshaar, B. and Martin, C. 2008. AtMYB12 regulates caffeoyl quinic acid and flavonol synthesis in tomato: Expression in fruit results in very high levels of both types of polyphenol. *Plant Journal* 56: 316–26.

Lurie, S., Shabtai, S. and Barg, R. 2003. Tomato plants and fruits with a transgenic *HSF* gene are more tolerant to temperature extremes. *Acta Horticulturae* 618: 201–7. doi: 10.17660/ActaHortic.2003.618.22.

Ma, H., Song, C., Borth, W., Sether, D., Melzer, M. and Hu, J. 2011. Modified expression of alternative oxidase in transgenic tomato and petunia affects the level of tomato spotted wilt virus resistance. *BMC Biotechnology* 11: 96.

Ma, J., Liu, T. and Qiu, D. 2015. Optimization of *Agrobacterium*-mediated transformation conditions for tomato (*Solanum lycopersicum* L.). *Plant Omics* 8: 529–36.

Ma, Y., Lin, S. Q., Gao, Y., Li, M., Luo, W. X., Zhang, J. and Xia, N. S. 2003. Expression of ORF2 partial gene of hepatitis E virus in tomatoes and immunoactivity of expression products. *World Journal of Gastroenterology* 9: 2211–15.

Mackenzie, K. F., Singh, K. K. and Brown, A. D. 1988. Water stress plating hypersensitivity of yeast: Protective role of trehalose in *Saccharomyces cerevisiae*. *Journal of General Microbiology* 134: 1661–6.

Mandaokar, A., Goyal, R., Shukla, A., Bisaria, S., Bhalla, R., Reddy, V., Chaurasia, A., Sharma, R., Altosaar, I. and Kumar, P. A. 2000. Transgenic tomato plants resistant to fruit borer (*Helicoverpa armigera* Hubner). *Crop Protection* 19: 307–12.

Mathews, H., Clendennen, S. K., Caldwell, C. G., Liu, X. L., Connors, K., Matheis, N., Schuster, D. K., Menasco, D. J., Wagoner, W., Lightner, J. and Wagner, D. R. 2003. Activation tagging in tomato identifies a transcriptional regulator of anthocyanin biosynthesis, modification, and transport. *Plant Cell Online* 15: 1689–703.

Mathieu-Rivet, E., Gevaudant, F., Sicard, A., Salar, S., Do, P. T., Mouras, A., Fernie, A. R., Gibon, Y., Rothan, C., Chevalier, C. and Hernould, M. 2010. Functional analysis of the anaphase promoting complex activator CCS52A highlights the crucial role of endo-reduplication for fruit growth in tomato. *Plant Journal* 62: 727–41.

Mattoo, A. K., Chung, S. H., Goyal, R. K., et al. 2007. Overaccumulation of higher polyamines in ripening transgenic tomato fruit revives metabolic memory, upregulates anabolism-related genes, and positively impacts nutritional quality. *Journal of AOAC International* 90: 1456–64.

Mattoo, A. K., Sobolev, A. P., Neelam, A., Goyal, R. K., Handa, A. K. and Segre, A. L. 2006. NMR spectroscopy-based metabolite profiling of transgenic tomato fruit engineered to accumulate spermidine and spermine reveals enhanced anabolic and nitrogen-carbon interactions. *Plant Physiology.* 142: 1759–70.

Mattoo, A. K. and Suttle, J. C. 1991. *The Plant Hormone Ethylene*. CRC Press, Boca Raton, FL.

Mattoo, A. K. 2014. Translational research in agricultural biology—enhancing crop resistivity against environmental stress alongside nutritional quality. *Frontiers in Chemistry* 2: 1–9. doi: 10.3389/fchem.2014.00030.

Mattoo, A. K., Upadhyay, R. K., and Ridrabhatla, S. 2015. Abiotic stress in crops: Candidate genes, osmolytes, polyamines and biotechnological intervention. In: Pandey, G. K. (ed.), *Elucidation of Abiotic Stress Signaling in Plants: A Functional Genomics Perspective*, Ch. 15. Springer Scientific and Business Media, New York, pp. 415–37.

McCormick, S., Niedermeyer, J., Fry, J., Barnason, A., Horsh, R. and Fraley, R. 1986. Leaf disc transformation of cultivated tomato (*L. esculentum*) using *Agrobacterium tumefaciens*. *Plant Cell Reports* 5: 81–4.

Medina, J., Catalá, R., and Salinasa, J. (2011). The CBFs: Three arabidopsis transcription factors to cold acclimate. *Plant Science* 180:3–11.

Mehta, R. A., Cassol, T., Li, N., Ali, N., Handa, A. K. and Mattoo, A. K. 2002. Engineered polyamine accumulation in tomato enhances phytonutrient content, juice quality, and vine life. *Nature Biotechnology* 20: 613–18.

Meissner, R., Jacobson, Y., Melamed, S., Levyatuv, S., Shalev, G., Ashri, A., Elkind, Y. and Levy, A. 1997. A new model system for tomato genetics. *Plant Journal* 12: 1465–72.

Melech-Bonfil, S. and Sessa, G. 2010. Tomato MAPKKKε is a positive regulator of cell-death signaling networks associated with plant immunity. *Plant Journal* 64: 379–91. doi: 10.1111/j.1365–313X.2010.04333.x

Meli, V. S., Ghosh, S., Prabha, T. N., Chakraborty, N., Chakraborty, S. and Datta, A. 2010. Enhancement of fruit shelf life by suppressing N-glycan processing enzymes. *Proceedings of the National Academy of Sciences USA* 107: 2413–18.

Metwali EMR., Soliman HIA., Fuller, M. P. and Almaghrabi, O. A. 2015. Improving fruit quality in tomato (*Lycopersicum esculentum* Mill) under heat stress by silencing the vis 1 gene using small interfering RNA technology. *Plant Cell, Tissue and Organ Culture* 121: 153–66.

Mittler, R., Vanderauwera, S., Gollery, M. and Van Breusegem, F. 2004. Reactive oxygen gene network of plants. *Trends in Plant Science* 9: 490–8.

Moghaieb, R., Saneoka, H. and Fujita, K. 2004. Shoot regeneration from Gus-transformed tomato (*Lycopersicon esculentum*) hairy root. *Cell and Molecular Biology Letters* 9: 439–49.

Moghaieb REA., Tanaka, N., Saneoka, H., Hussein, H. A., Yousef, S. S., Ewada, M. A. F., Aly, M. A. M. and Fujita, K. 2000. Expression of betaine aldehyde dehydrogenase gene in transgenic tomato hairy roots leads to the accumulation of glycine betaine and contributes to the maintenance of the osmotic potential under salt stress. *Soil Science and Plant Nutrition* 46(4): 873–83.

Monforte, A. J., Diaz, A., Caño-Delgado, A. and van der Knaap, E. 2014. The genetic basis of fruit morphology in horticultural crops: Lessons from tomato and melon. *Journal of Experimental Botany* 65: 4625–37.

Mooney, M., Desnos, T., Harrison, K., Jones, J., Carpenter, R. and Coen, E. 1995. Altered regulation of tomato and tobacco pigmentation genes caused by the *delila* gene of *Antirrhinum*. *Plant Journal* 7: 333–9.

Moyer, M. M., Londo, J., Gadoury, D. M., and Cadle-Davidson, L. (2015) Cold Stress-induced Disease Resistance (SIDR): Indirect effects of low temperatures on host-pathogen interactions and disease progress in the grapevine powdery mildew pathosystem. *Eur. J. Plant Pathol.* DOI: 10.1007/s10658-015-0745-1.

Muir, S. R., Collins, G. J., Robinson, S., Hughes, S., Bovy, A., Ric De Vos, C. H., van Tunen, A. J. and Verhoeyen, M. E. 2001. Overexpression of petunia chalcone isomerase in tomato results in fruit containing increased levels of flavonols. *Nature Biotechnology* 19: 470–4.

Muños, S., Ranc, N., Botton, E., et al. 2011. Increase in tomato locule number is controlled by two single-nucleotide polymorphisms located near WUSCHEL. *Plant Physiology* 156: 2244–54.

Murai, N., Sutton, D. W., Murray, M. G., et al. 1983. Phaseolin gene from bean is expressed after transfer to sunflower via tumor-inducing plasmid vectors. *Science* 222: 476–82.

Murlidhar Rao, M., Maruthi Rao, A., Kavikishor, P. B. and Jain, M. 2007. Thidiazuron enhanced shoot regeneration from different varieties of tomato (*Lycopersicon esculentum* Mill.). *Plant Cell Biotechnology and Molecular Biology* 8: 125–30.

Nambeesan, S., Datsenka, T., Ferruzzi, M. G., Malladi, A., Mattoo, A. K. and Handa, A. K. 2010. Overexpression of yeast spermidine synthase impacts ripening, senescence and decay symptoms in tomato. *Plant Journal* 63: 836–47.

Nath, P., Bouzayen, M., Mattoo, A. K. and Pech, J. C. 2014. *Fruit ripening: Physiology, Signalling and Genomics.* Wallingford: CAB International.

Neily, M. H., Matsukura, C., Maucourt, M., Bernillon, S., Deborde, C., Moing, A., et al. 2011. Enhanced polyamine accumulation alters carotenoid metabolism at the transcriptional level in tomato fruit over-expressing spermidine synthase. *Journal of Plant Physiology* 168(3): 242–52. doi:10.1016/j.jplph.2010.07.003.

Nicoletti, I., De Rossi, A., Giovinazzo, G. and Corradini, D. 2007. Identification and quantification of stilbenes in fruits of transgenic tomato plants (*Lycopersicon esculentum* Mill.) by reversed phase HPLC with photodiode array and mass spectrometry detection. *Journal of Agricultural and Food Chemistry* 55: 3304–11.

Oeller, P. W., Min-Wong, L., Taylor, L. P., Pike, D. A. and Theologis, A. 1991. Reversible inhibition of tomato fruit senescence by antisense RNA. *Science* 5030: 437–9.

Olías, R., Eljakaoui, Z., Li, J., De Morales PAZA., Marín-Manzano, M. C., Pardo, J. M. and Belver, A. 2009. The plasma membrane Na^+/H^+ antiporter SOS1 is essential for salt tolerance in tomato and affects the partitioning of Na^+ between plant organs. *Plant, Cell and Environment* 32: 904–16.

Orozco-Cardenas, M., McGurl, B. and Ryan, C. A. 1993. Expression of an antisense prosystemin gene in tomato plants reduces resistance toward *Manduca sexta* larvae. *Proceedings of the National Academy of Sciences USA* 90: 8273–6.

Ouyang, B., Chen, Y. H., Li, H. X., Qian, C. J., Huang, S. L. and Ye, Z. B. 2005. Transformation of tomatoes with osmotin and chitinase genes and their resistance to Fusarium wilt. *Journal of Horticultural Science and Biotechnology* 80: 517–22.

Pandey GK (ed.). 2015. Elucidation of Abiotic Stress Signaling in Plants. *Functional Genomics Perspectives*, Vol. 2. Springer, New York., p.488.

Pandey, S. K., Nookaraju, A., Upadhyaya, C. P., Gururani, M. A., Venkatesh, J., Kim D-H. and Park, S. W. 2011. An update on biotechnological approaches for improving abiotic stress tolerance in tomato. *Crop Science* 51: 2303–24.

Park, S., Li, J., Pittman, J. K., Berkowitz, G. A., Yang, H., Undurraga, S., et al. 2005. Up-regulation of a H+-pyrophosphatase (H+-PPase) as a strategy to engineer drought-resistant crop plants. *Proceedings of the National Academy of Sciences USA* 102: 18830–5.

Patade, V. Y., Khatri, D., Kumari, M., Grover, A., Mohan Gupta, S. and Ahmed, Z. 2013. Cold tolerance in Osmotin transgenic tomato (*Solanum lycopersicum* L.) is associated with modulation in transcript abundance of stress responsive genes. *Springer Plus* 2: 117.

Pramanik, M. H. R. and Imai, R. 2005. Functional identification of a trehalose-6-phosphatase gene that is involved in transient induction of trehalose biosynthesis during chilling stress in rice. *Plant Molecular Biology* 58: 751–62.

Purkayastha, S., Kaur, B., Arora, P., Bisyer, I., Dilbaghi, N. and Chaudhury, A. 2008. Molecular genotyping of *Macrophomina phaseolina* isolates: Comparison of microsatellite primed PCR and repetitive element sequence-based PCR *Journal of Phytopathology* 156: 372–81.

Purkayastha, S., Kaur, B., Dilbaghi N and Chaudhury, A. 2006. Characterization of *Macrophomina phaseolina*, the charcoal rot pathogen of cluster bean, using conventional techniques and PCR-based molecular markers. *Plant Pathology* 55: 106–16.

Rai, G. K., Rai, N. P., Kumar, S., et al. 2012. Effects of explant age, germination medium, pre-culture parameters, inoculation medium, pH, washing medium, and selection regime on *Agrobacterium*-mediated transformation of tomato. *In Vitro Cellular and Developmental Biology – Plant* 48: 565–78.

Raiola, A., Rigano, M. M., Calafiore, R., Frusciante, L. and Barone, A. 2014. Enhancing the human-promoting effects of tomato fruit for biofortified food. Hindawi Publishing Corporation Mediators of Inflammation.

Razdan, M. K. and Mattoo, A. K. 2006. *Genetic Improvement of Solanaceous Crops*: Volume 2: Tomato, Science Publishers, Inc., Enfield.

Rhim S.-L., Cho H.-J., Kim B-D., Schnetter, W. and Geider, K. 1995. Development of insect resistance in tomato plants expressing the δ-endotoxin gene of *Bacillus thuringiensis* subsp. *tenebrionis*. *Molecular Breeding* 1: 229–36.

Rodriguez, G. R., Muños, S., Anderson, C., Sim, S. C., Michel, A., Causse, M., Gardener, B. B. M., Francis, D., van der Knaap, E. 2011. Distribution of SUN, OVATE, LC, and FAS in the tomato germplasm and the relationship to fruit shape diversity. *Plant Physiology* 156: 275–85.

Römer, S., Fraser, P. D., Kiano, J. W., Shipton, C. A., Misawa, N., Schuch, W. and Bramley, P. M. 2000. Elevation of the provitamin A content of transgenic tomato plants. *Nature Biotechnology* 18, 666–9.

Ronen, G., Carmel-Goren, L., Zamir, D. and Hirschberg, J. 2000. An alternative pathway to β-carotene formation in plant chromoplasts discovered by map-based cloning of Beta and oldgold color mutations in tomato. *Proceedings of the National Academy of Sciences USA* 97: 11102–7.

Rosati, C., Aquilani, R., Dharmapuri, S., Pallara, P., Marusic, C., Tavazza, R., Bouvier, F., Camara, B. and Giuliano, G. 2000. Metabolic engineering of β-carotene and lycopene content in tomato fruit. *Plant Journal* 24: 413–20.

Roy, R., Purty, R. S., Agarwal, V. and Gupta, S. C. 2006. Transformation of tomato cultivar 'Pusa Ruby' with bspA gene from *Populus tremula* for drought tolerance. *Plant Cell, Tissue and Organ Culture* 84: 55–67.

Rus, A. M., Estan, M. T., Gisbert, C., et al. 2001. Expressing the yeast HAL1 gene in tomato increases fruit yield and enhances K+/Na+ selectivity under salt stress. *Plant, Cell and Environment* 24: 875–80. doi:10.1046/j.1365-3040.2001.00719.x.

Sanford, J. C., Chyi, Y. S. and Reisch, B. I. 1984. An attempt to induce egg transformation in *Lycopersicon esculentum* Mill. using irradiated pollen. *Theoretical and Applied Genetics* 67(6): 553–8.

Sato, S., Tabata, S., Hirakawa, H., Asamizu, E., Shirasawa, K., et al. 2012. The tomato genome sequence provides insights into fleshy fruit evolution. *Nature* 485: 635–41.

Schijlen, E., Ric de Vos, C. H., Jonker, H., Van Den Broeck, H., Molthoff, J., Van Tunen, A., Martens, S. and Bovy, A. 2006. Pathway engineering for healthy phytochemicals leading to the production of novel flavonoids in tomato fruit. Plant *Biotechnology Journal* 4: 433–44.

Schijlen EGWM., de Vos, C. H. R., Martens, S., Jonker, H. H., Rosin, F. M., Molthoff, J. W., Tikunov, Y. M., Angenent, G. C., van Tunen, A. J. and Bovy, A. G. 2007. RNA interference silencing of chalcone synthase, the first step in the flavonoid biosynthesis pathway, leads to parthenocarpic tomato fruits. *Plant Physiology* 144: 1520–30.

Schobert, B. 1977. Is there an osmotic regulatory mechanism in algae and higher plants? *Journal of Theoretical Biology* 68: 17–26.

Schreiber, G., Reuveni, M., Evenor, D., Oren-Shamir, M., Ovadia, R., Sapir-Mir, M., Bootbool-Man, A., Nahon, S., Shlomo, H., Chen, L. and Levin, I. 2012. *ANTHOCYANIN1* from *Solanum chilense* is more efficient in accumulating anthocyanin metabolites than its *Solanum lycopersicum* counterpart in association with the ANTHOCYANIN FRUIT phenotype of tomato. *Theoretical and Applied Genetics* 124: 295–307.

Seong, E. S., Baek, K.-H., Oh, S.-K., et al .2007. Induction of enhanced tolerance to cold stress and disease by overexpression of the pepper CaPIF1 gene in tomato. *Physiologia Plantarum* 129: 555–66. doi: 10.1111/j.1399-3054.2006.00839.x

Shen, H., Zhong, X., Zhao, F., et al. 2015. Overexpression of receptor-like kinase ERECTA improves thermotolerance in rice and tomato. *Nature Biotechnology* 33: 996–1003.

Shih C.-H., Chen, Y., Wang, M., Chu, I. K. and Lo, C. 2008. Accumulation of isoflavone genistin in transgenic tomato plants overexpressing a soybean isoflavone synthase gene. *Journal of Agricultural and Food Chemistry* 56: 5655–61.

Sigareva, M., Spivey, R., Willits, M., C. Kramer, C. and Chang, Y. F. 2004. An efficient mannose selection protocol for tomato that has no adverse effect on the ploidy level of transgenic plants. *Plant Cell Reports.* 23: 236–45.

Simkin, A. J., Gaffé, J., Alcaraz, J.-P., Carde, J.-P., Bramley, P. M., Fraser, P. D. and Kuntz, M. 2007. Fibrillin influence on plastid ultrastructure and pigment content in tomato fruit. *Phytochemistry* 68: 1545–56.

Simkin, A. J., Schwartz, S. H., Auldridge, M., Taylor, M. G. and Klee, H. J. 2004. The tomato carotenoid cleavage dioxygenase 1 genes contribute to the formation of the flavour volatiles β-ionone, pseudoionone, and geranylacetone. *Plant Journal* 40: 882–92.

Singer, M. A. and Lindquist, S. 1998. Thermotolerance in *Saccharomyces cerevisiae*: The yin and yang of trehalose. *Trends in Biotechnology* 16: 460–8.

Singh, N. K., Bracker, C. A., Hasegawa, P. M., Handa, A. K., Buckel, S., Hermodson, M. A. et al. 1987. Characterization of osmotin: A thaumatin-like protein associated with osmotic adaptation in plant cells. *Plant Physiology* 85: 529–36.

Singh, N. K., Handa, A. K., Hasegawa, P. M. and Bressan, R. A. 1985. Proteins associated with adaptation of cultured tobacco cells to NaCl. *Plant Physiology* 79: 126–37.

Smith, C. J. S., Watson, C. F., Morris, P. C., Bird, C. R., Seymour, G. B., Gray, J. E., et al. 1990. Inheritance and effect on ripening of antisense polygalacturonase genes in transgenic tomatoes. *Plant Molecular Biology* 1990 14: 369–79.

Sojikul, P., Buehner, N. and Mason, H. S. 2003. A plant signal peptide-hepatitis B surface antigen fusion protein with enhanced stability and immunogenicity expressed in plant cells. *Proceedings of the National Academy of Sciences USA* 100: 2209–14.

Soria-Guerra, E. R., Rosales-mendosa S., Marquez-Mercado, C., Lopez-Revilla, R., Castillo-Collazo, R., and Alpuche-Solis, G. A. 2007. Transgenic tomatoes express an antigenic polypeptide containing epitopes of the diphtheria, pertussis and tetanus exotoxins, encoded by a synthetic gene. *Plant Cell Reports* 26: 961–8.

Su, X., Xu, J., Rhodes, D., et al. 2016. Identification and quantification of anthocyanins in transgenic purple tomato. *Food Chemistry* 202: 184–8.

Sun, L., Yuan, B., Zhang, M., Wang, L., Cui, M., Wang, Q. and Leng, P. 2012. Fruit-specific RNAi-mediated suppression of *SlNCED1* increases both lycopene and β-carotene contents in tomato fruit. *Journal of Experimental Botany* 63: 3097–108.

Sun, Y., Dilkes, B. P., Zhang, C., Dante, R. A., Carneiro, N. P., Lowe, K. S., Jung, R., Gordon-Kamm, W. J. and Larkins, B. A. 1999. Characterization of maize (*Zea mays* L.) Wee1 and its activity in developing endosperm. *Proceedings of the National Academy of Sciences USA* 96: 4180–5.

Swindell, W. R., Huebner, M. and Weber, A. P. 2007. Transcriptional profiling of *Arabidopsis* heat shock proteins and transcription factors reveals extensive overlap between heat and non-heat stress response pathways. *BMC Genomics* 8: 125.

Tabaeizadeh, Z., Agharbaouri, Z., Harrak, H. and Poysa, V. 1999. Transgenic tomato plants expressing a *Lycopersicon chilense* chitinase gene demonstrate improved resistance to *Verticillium dahlia* race 2, *Plant Cell Reports* 19: 197–202.

Tackaberry, E. S., Prior, F., Bell, M., Tocchi, M., Porter, S., Mehic, J., Ganz, P. R., Sardana, R., Altosaar, I. and Dudani, A. 2003. Increased yield of heterologous viral glycoprotein in the seeds of homozygous transgenic tobacco plants cultivated underground. *Genome* 46: 512–26.

Tang, X. Y., Xie, M. T., Kim, Y. J., Zhou, J. M., Klessig, D. F. and Martin, G. B. 1999. Overexpression of Pto activates defense responses and confers broad resistance. *Plant Cell* 1: 15–29.

Thakur, B. R., Singh, R. K. and Handa, A. K. 1996a. Effect of an antisense pectin methylesterase gene on the chemistry of pectin in tomato (*Lycopersicon esculentum*) juice. *Journal of Agricultural and Food Chemistry* 44: 628–30.

Thakur, B. R., Singh, R. K., Tieman, D. M. and Handa, A. K. 1996b. Tomato product quality from transgenic fruits with reduced pectin methylesterase. *Journal of Food Science* 61: 85–7.

Thipyapong, P., Hunt, M. D. and Steffens, J. C. 2004. Anti sense downregulation of polyphenol oxidase results in enhanced disease susceptibility. *Planta* 220: 105–7.

Thomashow, M. F. 1999. Plant cold acclimation: Freezing tolerance genes and regulatory mechanisms. *Annual Review of Plant Physiology and Plant Molecular Biology* 50: 571–99.

Thomashow, M. F. (2010). Molecular basis of plant cold acclimation: Insights gained from studying the CBF cold response pathway. *Plant Physiology* 154:571–7.

Tieman, D., Taylor, M., Schauer, N., Fernie, A. R., Hanson, A. D. and Klee, H. J. 2006 Tomato aromatic amino acid decarboxylases participate in synthesis of the flavour volatiles 2-phenylethanol and 2-phenylacetaldehyde. *Proceedings of the National Academy of Sciences USA* 103: 8287–92.

Tieman, D., Zeigler, M., Schmelz, E., Taylor, MG., Rushing, S., Jones, J. B. and Klee, H. J. 2010. Functional analysis of a tomato salicylic acid methyl transferase and its role in synthesis of the flavour volatile methyl salicylate. *Plant Journal* 62: 113–23.

Tieman, D. M. and Handa, A. K. 1994. Reduction in pectin methylesterase activity modifies tissue integrity and cation levels in ripening tomato (*Lycopersicon esculentum* Mill.) fruits. *Plant Physiology* 106: 429–36.

Tieman, D. M., Harriman, R. W., Ramamohan, G. and Handa, A. K. 1992. An antisense pectin methylesterase gene alters pectin chemistry and soluble solids in tomato fruit. *The Plant Cell* 4: 667–79.

Ume-e-Ammara, Al-Maskri, A. Y., Khan, A. J., Al-Sadi, A. M. 2014. Enhanced somatic embryogenesis and *Agrobacterium*-mediated transformation of three cultivars of tomato by exogenous application of putrescine. *International Journal of Agriculture and Biology* 16: 81–8.

Upadhyay, R. K., Gupta, A., Ranjan, S., Singh, R., Pathre U V., Nath, P. and Sane, A. P. 2014. The EAR Motif Controls the Early Flowering and Senescence Phenotype Mediated by Over-Expression of SlERF36 and Is Partly Responsible for Changes in Stomatal Density and Photosynthesis In: Meyer, P. (ed.), PLoS ONE 9.

Upadhyay, R. K., Soni, D. K., Singh, R., Dwivedi, U. N., Pathre U V., Nath, P. and Sane, A. P. 2013. SlERF36, an EAR-motif-containing ERF gene from tomato, alters stomatal density and modulates photosynthesis and growth. *Journal of Experimental Botany* 64: 3237–47.

Van der Wel, H. and Loeve, K. 1972. Isolation and characterization of thaumatin I and thaumatin II the sweet tasting proteins from *Thaumatococcus daniellii* Benth. *European Journal of Biochemistry* 31: 221–5.

Ververidis, F., Trantas, E., Douglas, C., Vollmer, G., Kretzschmar, G. and Panopoulos, N. 2007. Biotechnology of flavonoids and other phenylpropanoid-derived natural products. Part II: Reconstruction of multienzyme pathways in plants and microbes. *Biotechnology Journal* 2: 1235–49.

Virginia Polytechnic Institute and State University. Information Systems for Biotechnology, Field Test Releases in the US. 2016. http://www.isb.vt.edu.

Wang, J. Y., Lai, L. D., Tong, S. M. and Li, Q. L. 2013. Constitutive and salt-inducible expression of SlBADH gene in transgenic tomato (*Solanum lycopersicum* L. cv. Micro-Tom) enhances salt tolerance. *Biochemical and Biophysical Research Communications* 432: 262–7. doi:http://dx.doi.org/10.1016/j.bbrc.2013.02.001.

Wang, L., Zhao, Y., Reiter, R. J., He, C., Liu, G., Lei, Q., Zuo, B., Zheng, X. D., Li, Q. and Kong, J. 2014. Changes in melatonin levels in transgenic "Micro-Tom" tomato overexpressing ovine AANAT and ovine HIOMT genes. *Journal of Pineal Research* 56(2): 134–42. doi:10.1111/jpi.12105.

Wang, S., Liu, J., Feng, Y., Niu, X., Giovannoni, J. and Liu, Y. 2008. Altered plastid levels and potential for improved fruit nutrient content by downregulation of the tomato DDB1-interacting protein CUL4. *Plant Journal* 55: 89–103.

Wang, S. F., Wang, J. L., Zao, Y. S. and Zhang, H. 2001. Transformation of choline dehydrogenease gene and identification of salt-tolerance in transgenic tomato. *Acta Phytophysiologica Sinica* 27(3): 248–52.

Wang, W., Vinocur, B. and Altman, A. 2003. Plant responses to drought, salinity and extreme temperatures: Towards genetic engineering for stress tolerance. *Planta* 218:1–14.

Warzecha, H. and Mason, H. S. 2003. Benefits and risks of antibody and vaccine production in transgenic plants. *Journal of plant physiology* 160: 755–64.

Wu, Z., Sun, S., Wang, F. and Guo, D. 2011a. Establishment of regeneration and transformation system of *Lycopersicon esculentum* Micro tom. British *Biotechnology Journal* 1:53–60.

Wu, S., Xiao, H., Cabrera, A., Meulia, T., van der Knaap, E. 2011b. SUN regulates vegetative and reproductive organ shape by changing cell division patterns. *Plant Physiology* 157: 1175–86.

Wurbs, D., Ruf, S. and Bock, R. 2007. Contained metabolic engineering in tomatoes by expression of carotenoid biosynthesis genes from the plastid genome. *Plant Journal* 49: 276–88.

www.lib.purdue.edu

www.nationalacademies.org. Genetically engineered crops: experiences and prospects. May 2016.

Xiao, H., Jiang, N., Schaffner, E., Stockinger, E. J., van der Knaap, E. 2008. A retrotransposon mediated gene duplication underlies morphological variation of tomato fruit. *Science* 319: 1527–30.

Xie, Q., Liu, Z., Meir, S., Rogachev, I., Aharoni, A., Klee, H. J. and Galili, G. 2016. Altered metabolite accumulation in tomato fruits by co-expressing a feedback-insensitive *AroG* and the *PhODO1 MYB-type* transcription factor. Plant *Biotechnology Journal*. doi:10.1111/pbi.12583 [Epub ahead of print].

Yang, J. C., Zhang, J. H., Liu, K., Wang, Z. Q. and Liu, L. J. 2007. Involvement of polyamines in the drought resistance of rice. *Journal of Experimental Botany* 58: 1545–55.

Yang, S., Gao, L., Sun, X., Li, H., Deng, H. and Liu, Y. 2015. Over-expressing SlWD6 gene to improve drought and salt tolerance of tomato. *Chinese Journal of Applied and Environmental Biology* 3: 413–20.

Yasmeen, A., Mirza, B., Inayatullah, S., Safdar, N., Jamil, M., Ali, S. and Choudry, M. F. 2009. *In planta* transformation of tomato. *Plant Molecular Biology Reports* 27:20–8.

Ye, X., Al-Babili, S., Kloti, A., Zhang, J., Lucca, P., Beyer, P. and Potrykus, I. 2000. Engineering the provitamin A (beta-carotene) biosynthetic pathway into (carotenoid-free) rice endosperm. *Science* 287: 303–5.

Youm, J. W., Jeon, J. H., Kim, H., Kim, Y. H., Ko, K., Joung, H. and Kim, H. S. 2008. Transgenic tomatoes expressing human beta-amyloid for use as a vaccine against Alzheimer's disease. *Biotechnology Letters* 30: 1839–45.

Youm, J. W., Kim, H., Han, J. H. L., Jang, C. H., Ha, H. J., Mook-Jung, I., Jeon, J. H., Choi, C. Y., Kim, Y. H., Kim, H. S. and Joung, H. 2005. Transgenic potato expressing Ab reduce Ab burden in Alzheimer's disease mouse model. *FEBS Lett* 579: 6737–44.

Yu, S., Wang, W. and Wang, B. 2012. Recent progress of salinity tolerance research in plants. *Russian Journal of Genetics* 48: 497–505.

Zhang, H. and Blumwald, E. 2001. Transgenic salt-tolerant tomato plants accumulate salt in foliage but not in fruit. *Nature Biotechnology* 19: 765–8.

Zhang, Y., Butelli, E., Alseekh, S., Tohge, T., Rallapalli, G., Luo, J., Kawar, P. G., Hill, L., Santino, A.,Fernie, A. R. and Martin, C. 2015. Multi-level engineering facilitates the production of phenylpropanoid compounds in tomato. *Nature Communications* 6: 8635.

Zhao, J.-Z., Cao, J., Li, Y., Collins, H. L., Roush, R. T., Earle, E. D. and Shelton, A. M. 2003. Transgenic plants expressing two Bacillus thuringiensis toxins delay insect resistance evolution. *Nature Biotechnology* 21: 1493–7.

Zhu, J.-K., Liu, J. and Xiong, L. 1998. Genetic analysis of salt tolerance in *Arabidopsis*: Evidence for a critical role of potassium nutrition. *Plant Cell* 10: 1181–91.

Zhu, J.-K. 2002. Salt and drought stress signal transduction in plants. *Annual Review of Plant Biology* 53: 247–73. doi:10.1146/annurev.arplant.53.091401.143329.

Zhu, Q. and Berzofsky, J. A. 2013. Oral vaccines directed safe passage to the front line of defense. *Gut Microbes* 4: 246–52.

Lightning Source UK Ltd.
Milton Keynes UK
UKHW021312030221
378174UK00006B/51